WATER MANAGEMENT AND VIOLENT CONFLICT IN EAST AFRICA

This book examines the complex interrelationships between water availability, governance and violent and non-violent conflicts, drawing on in-depth case studies of Lake Naivasha in Kenya and Lake Wamala in Uganda.

When international economic endeavours like flower farming, oil exploration and extensive rice growing are coupled with a government's prioritization of economic development, not only does this lead to the commercialization of water resources but it also creates conflicts between national, regional and local stakeholders. Often overlooked in existing literature, such is the case even in water abundant areas like Lake Naivasha and Lake Wamala. Presenting a comparative study, the book provides a unique perspective on multifunctional water use and illustrates how politically and economically motivated water use increases violent tensions over access to and the use of freshwater resources. The coverage stretches from international and national agencies to NGOs, economic stakeholders and local actors. The book evaluates the resilience and vulnerability of local actors ability to access water and examines the nexus between the need to access water and the ability to influence access to water, taking into consideration both countries' economic development agendas. The book concludes by discussing strategies for reducing water-induced conflicts that can be applied to not only these cases but water conflicts across the globe.

This book will be of great interest to scholars and professionals of water resource management and governance, African development, conflict resolution and sustainable development.

Julia Renner-Mugono (PhD) is Social and Environmental Specialist for East Africa and the African Union at a German Development Institution and Associate Fellow at the Peace Academy of Rhineland-Palatinate, Germany.

Earthscan Studies in Water Resource Management

Water Management in China's Power Sector
Xiawei Liao and Jim W. Hall

Reviving Indigenous Water Management Practices in Morocco
Alternative Pathways to Sustainable Development
Sandrine Simon

Transnational Corporations in Urban Water Governance
Public – Private Partnerships in Mexico and the US
Joyce Valdovinos

Defining Effective Transboundary Water Cooperation
Melissa McCracken

Corporatization and the Right to Water in Colombia
Conflicts, Citizenship and Social Inequality
Marcela López

Gender Dynamics in Transboundary Water Governance
Feminist Perspectives on Water Conflict and Cooperation
Edited by Jenniver Sehring, Rozemarijn ter Horst, and Margreet Zwarteveen

Water Management and Violent Conflict in East Africa
Scarcity and Security in Kenya and Uganda
Julia Renner-Mugono

For more information about this series, please visit: www.routledge.com/books/series/ECWRM/

WATER MANAGEMENT AND VIOLENT CONFLICT IN EAST AFRICA

Scarcity and Security in Kenya and Uganda

Julia Renner-Mugono

Routledge
Taylor & Francis Group
LONDON AND NEW YORK

earthscan
from Routledge

Designed cover image: © Getty Images

First published 2023
by Routledge
4 Park Square, Milton Park, Abingdon, Oxon OX14 4RN

and by Routledge
605 Third Avenue, New York, NY 10158

Routledge is an imprint of the Taylor & Francis Group, an informa business

British Library Cataloguing-in-Publication Data
A catalogue record for this book is available from the British Library

ISBN: 978-1-032-22144-1 (hbk)
ISBN: 978-1-032-22145-8 (pbk)
ISBN: 978-1-003-27129-1 (ebk)

DOI: 10.4324/9781003271291

Typeset in Bembo
by Apex CoVantage, LLC

To my grandparents Christel and Eberhard Scharf.

CONTENTS

FIGURES

TABLES

ABBREVIATIONS

ACLED	Armed Conflict Location and Event Dataset
CGoN	County Government of Nakuru
DRC	Democratic Republic of Congo
EU	European Union
FAO	Food and Agricultural Organization
GDP	Gross Domestic Product
GoK	Government of Kenya
ICRC	International Committee of the Red Cross
ICT	Information and Communication Technology
IPCC	International Panel on Climate Change
LNGG	Lake Naivasha Growers Group
LNRA	Lake Naivasha Riparian Association
LNRUA	Lake Naivasha Resource Users Association
MTP	Medium-Term Plan
MWE	Ministry for Water and Environment
NDP	National Development Plan
NEMA	National Environmental Management Authority
NGO	Non-governmental Organization
PRIO	Peace Research Institute Oslo
SA	Stakeholder Analysis
SCAD	Social Conflict Analysis Database
UCDP	Uppsala Conflict Data Program, Resource Conflict Institute
UNDESA	United Nations Department for Economic and Social Affairs
UNFCCC	United Nations Framework Convention on Climate Change
WHO	World Health Organization
WWF	World Wildlife Fund

1

INTRODUCTION

Water is treated as an infinite resource as it is a prerequisite for livelihood and survival. However, increasing demand and simultaneously decreasing supply are aggravating the pressures on this treasured resource. Additionally, it is also getting warmer. During the past century, the global temperature has increased by 0.85 degrees Celsius (IPCC 2014). In this context, the *IPCC Special Report: Global Warming of 1.5°C* highlights the risk of increased flooding and prolonged droughts if the effects of climate change are not mitigated (IPCC 2018a). Across the scientific community, a broad consensus has been reached that socio-economic conditions have a greater impact on the available water resources and associated risk than the changes in climate (Gabzdylova et al. 2009; Ali et al. 2017). Anthropogenic activities, especially economic investments and rapid population growth, have resulted in land cover changes, deforestations and wetland destructions. Taken together, these actions reduce the water storage capacity of trees and other plants. Plants and trees are responsible for extracting groundwater from soil and returning it into the atmosphere. Thus, this alters the water cycle and leads to higher temperatures, desertification and lower rainfall, especially during the dry season (Vergopolan and Fisher 2016; Sumila et al. 2017). The international community has acknowledged that climatic and non-climatic drivers can have severe effects on human livelihoods across the globe (IPCC 2018b).

Considering the population and economic growth rates of many developing countries, it seems unlikely that the continuation of global warming below 2 degrees Celsius above pre-industrial levels can be prevented (UNFCCC 2015). Current trajectories show that if global warming is not constrained to less than 2°C, 8 per cent of the contemporary population (7.7 billion) is exposed to a severe reduction in water resources till 2040 and 14 per cent of the population by 2071 (Gerten et al. 2013; Schewe et al. 2014). While those are extreme scenarios, it is widely recognized that climate change has the potential to undermine the adaptive

DOI: 10.4324/9781003271291-1

capacity of vulnerable populations (IPCC 2014, 2018a). Further, concerns have been raised that changes due to climate change will aggravate water-related hazards and water scarcity (UNFCCC 2011). This has the potential to undermine human security (UNDESA 2014). The underlying assumption is that both raising temperatures and changed precipitation reduce the availability of water resources, which, in turn, intensifies tensions among the resource users. Furthermore, in areas and regions that are particularly vulnerable and already affected by conflicts, tensions are assumed to intensify over the usage and accessibility of water resources (see, e.g. IPCC 2007, 2014, 2018b; Kummu et al. 2016; O´Connell 2017; European Commission 2020).

Many African states which are either vulnerable to water shortages (East Africa), water stress (South and North-West Africa) or water scarcity (North Africa), have a history of recent conflicts (since 2010) which involve water as a subject or face a double exposure to any form of water shortages and conflict. The causes of the conflict are highly diverse but, generally, the continent's vulnerability to water stress can be attributed to its strong exposure to climate change and its limited adaptive capacity (Busby et al. 2012; IPCC 2014, 2018a). Additionally, the African population is sensitive to water shortages because of its dependence on rain-fed agriculture for income and food security (FAO 2020). Corruption, poverty and low levels of education put additional stress on the population's adaptive capacity (Transparency International 2020; World Bank Group 2020).

A hotspot mapping confirms that most of the reported water conflicts take place in the arid areas of East Africa (e.g. Somalia, Sudan, South Sudan, Ethiopia, Northern Kenya and North-Eastern Uganda) (IPCC 2018a; Pacific Institute 2020). This has caught the attention of the scientific community who discussed the relationship between changing climate on water resources and weather, and if so, how this increases the risk of violent conflict (see, e.g. Dinar et al. 2011; Gleditsch 2012; Hisang et al. 2013 or Almer et al. 2017). There appears to be consensus that there is a positive relationship between increases in temperature and increases in human conflict. However, the depletion of natural resources as a result of overexploitation, rapid population growth or economic and social interest in water resources accentuates the need to pay attention to regions and areas which do not seem water stressed or water scarce.

1.1 Book's Research Objective

Amid the background of intensifying competition over water resources, in addition to and beyond the stressor of climate change, it becomes increasingly important to understand the effects of social, economic, political and cultural factors of the outbreak of conflicts. This includes countries and regions which have not been affected by water shortages as a result of climate change yet. The effects of the aforementioned factors can only be understood if the different aspects of vulnerability and adaptive capacity for a variety of stakeholders are analysed first. The majority of research has concentrated on water stressed and water scarce areas and its interrelation with one or two of the previously mentioned variables, including the aspect of climate change. Additionally, many publications focused on cross-border water basins (e.g. Nile River basin) which blurs the variety of interconnected causes for the outbreak of water-related conflicts. Even if the population of Africa doubles by 2050 (UNDESA 2020) and the potential for internal migration rises as a consequence of extreme weather events and the drying up of water bodies, the majority of the country's population will move towards internal, still water-rich bodies. Furthermore, the wish of African governments to achieve middle-income status by 2050 and the growing complexity of international and powerful national companies will have important implications on the availability of natural resources. Against this backdrop, this book seeks to shed light on the causes and dynamics of water-related conflicts at inner-state water bodies which are located in humid and wetland areas.

Up to now, it is the first time that the relationship between political, economic and social forms of vulnerability, adaptive capacity and non-violent and violent forms of conflict are discussed and analysed jointly covering two inner state lakes which seem both calm on the surface and well-endowed with water. The research region and sites, Lake Naivasha in Kenya and Lake Wamala in Uganda, were chosen to cover a variety of international, national and local stakeholders. Moreover, the lake sites cover a variety of attributes associated with different levels of economic development, advanced decentralization of political processes, population growth and composition as well as conflict type and behaviour.

The analysed conflict types include both non-violent and violent behaviour without state involvement. At Lake Naivasha, the focus is placed on small-holder farmer and pastoralist conflicts, whereas at Lake Wamala, the small-holder farmer and fishermen conflict is analysed in more detail. Both types of conflict have a strong resource component because the actors depend on water and land to secure their daily livelihood. These forms of non-violent and violent behaviour often occur in arid and semi-arid lands which cover around 41 per cent of the world's surface (United Nations 2020). Globally, around 1.5 billion people depend directly on these degrading areas. However, with global climate change, it is expected that arid and semi-arid lands will increase and, therewith, human activities centre on humid and sub-humid areas (Ibid. 2020). Hence, the findings of this study are likely to be relevant beyond the chosen lake sites. Consequently, the research findings from this book contribute to the emerging knowledge of the role of political, economic and social factors of vulnerabilities and actors influencing the governance of water resources on non-violent and violent conflicts in water abundant areas. Using the case studies of Lake Naivasha and Lake Wamala, this book is guided by the overarching question: "How does water management influence inner state low-key conflict intensity in Kenya and Uganda – with a specific focus on Lake Naivasha in Kenya and Lake Wamala in Uganda?"

To grasp the complexity of this research aim, a multitude of research methods needed to be applied at different geographical scales and across cultural and political boundaries (see section 1.3). To satisfy the interdisciplinarity, the book combines a political science approach with perspectives from different disciplines, including peace and security studies, economics and sociology, as well as ecology and agricultural sciences. To ensure a comprehensive local perspective, field research was conducted trice. During this time, journeys to the lake sites took place to gain first-hand information about the water and conflict nexus. Furthermore, the author lived with the different conflict parties at Lake Naivasha and Lake Wamala, partly under challenging security and climatic conditions.

In the course of the book, some main terms are used frequently. Among others, the term 'natural resources' is used to describe both land and water resources if not specified otherwise. The name stakeholder and actor are applied interchangeably to refer to the different groups of actors or individuals who are involved in water-related conflicts. The term 'community' characterizes a group of people who live in a village or landing site, either at the lake or in its closer surroundings, and directly depend on the lake's resources. Water body refers to either lakes in general or the lake sites studied. The term 'conflict' is used to describe non-violent and violent tensions which erupt between the different groups of actors as a result of incompatible interests. Additional terms used will be defined according to their occurrence.

1.2 Research Area: A Brief Introduction

Located in East Africa, Kenya and Uganda are well-endowed with internal and cross-border lakes.[1] In total, there are 9 larger lakes in Kenya and 15 in Uganda. For Kenya, 97.7 per cent of the water bodies are permanent, whereas for Uganda, 95.5 per cent lakes are permanent water bodies. Permanent water is defined as being present for all 12 months a year. Since 2000, the extent of permanent and seasonal surface water has increased by 2.64 per cent for Kenya. Uganda denoted a loss of permanent and seasonal surface water by 2.64 per cent. Furthermore, 1 out of 9 lakes (Lake Turkana in Kenya) and 2 out of 15 lakes (Lake Victoria and Lake Albert in Uganda) are classified as being in a state of extreme water quality. The observed gains and losses in permanent water levels do not apply to all lakes equally. While the water levels of the lakes which are located in the arid areas reduced in their sizes (i.e. Lake Turkana in Kenya and Lake Opeta and Lake Kyoga in Uganda), lakes in humid and wetland areas increased in their water levels – at least temporarily (European Commission 2020). The water quality for Lake Naivasha and Lake Wamala is categorized as normal to low. Beyond the mentioned ecological aspects, Lake Naivasha and Lake Wamala have been accessible also in terms of safety and, consequently, qualified as case studies.[2]

1.3 Methodology Used

The research strategy adopted is based on qualitative and interpretive social research. The term 'interpretive research' or 'interpretive methods' was introduced by Thomas Wilson (1970) to make an explicit distinction between normative and interpretive paradigms. He argues that instead of encountering and reacting to the world humans find themselves placed within, they create a social reality through continuous interactions with other human beings (Wilson 1970). The researcher uses open methods including various forms of open, semi-structured and structured interviews, field research consisting of participatory and non-participatory observations, and audio and video recordings of everyday situations, group discussions or community interviews (Rosenthal 2018). The aim is to describe the research object from the perspective of the 'actors in everyday contexts . . . and to study complex social actions and practices of everyday situations' (Ibid. 2018: 15). Interpretive social research provides information about how people construct their social reality. Both underlying typical (i.e. regularly occurring) sense structures and latent meanings of behaviour can be reconstructed using an empirical case-by-case analyses (Kleemann et al. 2009). Therefore, in this context, scholars often refer to 'explorative' (i.e. discovering or exploring) procedures or to the 'principle of openness' to the object of investigation (Rosenthal 2018).

Thus, a central basis of interpretive methods is a strictly sequential analysis method, which determines the exact order of the linguistic and non-linguistic activities. Therewith, behaviour and observations are central elements to detect meaning. For interpretive social research, all components of statements have a meaning, including spoken ones, written texts but also mental processes and physical actions. The interpretation of the text and, therewith, the reconstruction of its overall social meaning is key to an implicit knowledge generation (Kleemann et al. 2009; Rosenthal 2018). The phrase *text* is used broadly and 'refer[s] to all expressive configurations produced during a social interaction which have been recorded in some way or [an]other' (Rosenthal 2018: 18). The textual approach analyses the interviews regarding their content which is generated by the answers of interviewees to the prepared questions. The analysis of the text follows a structured process and each sequence is considered a part of the whole text. This differentiation makes it possible to examine not only the manifest content of the text but also its latent meaning. Contrary to most content analysis approaches, qualitative and interpretive social research provides a tool to discover the meaning 'that is hidden between the lines' (Rosenthal 2018: 17). Thus, interpretive methods which seem to be hidden at first are suitable to uncover hidden layers of meanings. This enables to make the different layers from the surface visible 'in a manner that is methodologically controlled and intersubjectively testable' (Hitzler and Honer 1997: 23).

The empirical study begins with a very broad research question, which is concretized and refined across the research process and subdivided into more detailed research questions. Hence, there are no predefined hypotheses (Rosenthal 2018). Instead of merely testing hypotheses empirically in the course of data analysis, researchers should always be open to discovering new connections or other constellations of known factors in interpretive data analysis (Kleemann et al. 2009). Strauss and Corbin (1990) specify the principle of openness and argue that at the beginning of an empirical study, open questions 'indicate what the research is to be focused on and what the researcher wants to know about this object' (Strauss and

Corbin 1990: 37). The general question of the investigation within the framework of interpretive social research should therefore be formulated in such a way that it does not contain any content-related assumptions or explanatory hypotheses on the object of investigation. Rather, the researcher uses methods of formal theoretical insight to research empirical cases to obtain knowledge beyond a concrete reference to an object (Kleemann et al. 2009). The aim is not so much to review existing theories as it is either to discover new phenomena from existing studies or to further develop them either theoretically or empirically. The research question and, therewith, the main subject to investigate was formulated on the basis of a literature review and existing theories. In the course of this process, the research question was subdivided into concrete sub-research questions, which provide the foundation to answer the overall research question posed in this book (Döring and Bortz 2016). The research question was answered through qualitative structured, semi-structured and open interviews and observations. Accordingly, the study is question based instead of hypotheses based.

The method of interpretive social research has suffered, and to some extent still does suffer from a number of misconceptions about its worth as a scientific research method. Even though this method is particularly suitable for reconstructing latent meanings, there remain some limitations. The difference between intended and objective meanings applies also to a large extent to the undertaken actions and the knowledge of the researcher itself. While this is theoretically possible during the time of action or during the time of reflecting on the conducted interviews and undertaken observations, researchers, however, remain subject to the situation that create this difference, respectively. As implicit knowledge is key during the researchers' action in their research process, it can never be fully revealed or reflected on (Rosenthal 2018). Furthermore, certain meanings of every text might not become obvious to the researcher given, on the one hand, the different socializations of the researcher and the research objects within a specific social and historical context and, on the other hand, the discourses and the social circumstances by which these are influenced. This, therefore, might lead to misinterpretations, particularly concerning how historical situations impact the present situation.

1.3.1 Case Study Methodology

Qualitative and interpretive social research focuses on a small number of cases 'that are expected to provide insight into a causal relationship across a larger population of cases' (Gerring 2007: 86). Furthermore, this method is used to not only develop but also criticize diverse theories based on empirical inquiry. Thus, a case study methodology is a suitable research strategy for this book which can be adopted as a subgrouping of qualitative and interpretive social research.

Yin (2003) defines case study as 'an empirical inquiry that investigates a contemporary phenomenon within its real-life context when the boundaries between phenomenon and context are not clearly evident' (Yin 2003: 13). *Case* refers to a 'spatially delimited phenomenon (a unit) observed at a single point in time or over some period of time' (Gerring 2010). A case study attempts to elucidate certain features from nation-states or the behaviour of individuals, for example. Thus, it comprises either single or multiple (within-case) observations (Odell 2001). Case study methodology might incorporate one single case study but can apply to several cases in the form of a multiple case study as well. As the number of cases to be investigated increases, so too does the likelihood of intense individual case investigation decrease. Therefore, the fewer cases there are, and 'the more intensively they are studied, the more a work merits the appellation case study' (Gerring 2010: 20).

Another common type of case study analysis aims to illustrate a theory (Odell 2001). Either the development or the testing of already-developed theories prior to the empirical data collection, and analysis thereof, is an essential part of case study inquiries. The theory or different theoretical strands put an abstract idea 'in order to help readers see its meaning more clearly, and to convince them that the idea is relevant to at least one significant real-world instance' (Odell 2001: 163). Often, this method relies on multiple sources of evidence to increase the reliability of the findings through, for example, a triangulation of data (Gerring 2010). Similar to qualitative and interpretive social research, but more precise, the preferred strategy when conducting case studies is to ask *how* and *why* types of research questions concerning contemporary events over which the research has either little or no control (Yin 2003). How and why types of questions allow the interviewee to share their experience in depth, giving them freedom to express their thoughts. Therefore, this method is contrary to surveys, experiments or archival analysis.

The case study methodology as a subcategory of qualitative and interpretive social research was selected to answer the research questions presented in chapter 1 of this book. Since this method is more concrete than qualitative and interpretive social research, case study methodology presents a crucial tool to understand complex social, organizational and political phenomena. These phenomena usually unfold in practice through approaching the real-life situation (Flyvbjerg 2001, 2006; Yin 2003). As already touched upon, this approach relies on several different sources of information, including structured and semi-structured interviews and focus group discussions, documentary information and participatory and non-participatory observations of the research objects. The chosen sources allow the researcher to 'develop converging lines of inquiry' (Yin 2003: 98) in order to triangulate data and strengthen the general quality as well as reliability of the case study findings (Ibid. 2003).

Similar to qualitative and interpretive social research, the case study approach has suffered and still does suffer from a number of impartial ideas and misconceptions about its value as a scientific research method. Flyvbjerg (2006) captured some misperceptions in his article *Five Misunderstandings About Case-Study Research*. Among other misunderstandings, he includes the notions that 'general theoretical (context-independent) knowledge is more valuable than concrete practical (context-dependent) knowledge' (Flyvbjerg 2006: 221), that cannot be generalized on the basis of an individual case and that case studies, therefore, tend to confirm the researchers' perceived ideas (Ibid. 2006). However, this critique has largely been refuted. Nevertheless, and compared to other methods and strategies, case study research is still criticized for its lack of a consistent systematic approach regarding the implementation and conduction of the case study. This allegation has led to a supposed shortage of rigor case study research (Yin 2003).

A consequence of this lack of rigor is that there is no conclusive evidence that may affect the direction of case study findings (Yin 2003), which might result in the fact that the case study selection seems to come across as being self-evident (Lund 2014). In order to relativize this criticism, Lund (2014) further argues that scholars '"see" things through a lens of *a priory* concepts formed through experience' (Lund 2014: 226). The case study concept is charged for not being objective as they guide the researchers' inquiry by a closer examination of certain dynamics, processes and relations, while occluding others. In case, the concept through which the researchers understand the world and orient the inquiry is not called into question, one easily finds oneself in a situation where observations are judged as being (mistakenly) unequivocal and pervasive. To avoid this shortcoming, a detailed examination of possible frames of inquiry, including the choice of concepts as well as theoretical questions, is indispensable. Moreover, the considered concepts are often quite abstract, even to an extent that these do not seem to have any immediate empirical reference. Therefore, it is necessary to track and explain the movement of any taken step and to be explicit about how these concepts are operationalized into observable, empirical elements (Lund 2014).

Based on Lund's definition, 'things' are looked at through the lens of the IPCC's Vulnerability Concept, Stakeholder Analysis (SA) and Conflict Analysis as well as their related concepts as discussed in the previous chapter. As mentioned in the theoretical chapter, Methmann and Oels (2014), Grimble and Wellard (1997) as well as Ogada et al. (2017) ask in these theoretical approaches if the Vulnerability Concept and SA are more methodological approaches for an identification and classification of the interests and influences of particular actors in relation to natural resource governance processes rather than a theory. In a similar vein, within his systemic thinking conflict approach, Ropers (2008) distinguishes between a comprehensive conflict analysis, which is heuristic (i.e. it allows for a better description of the conflict's context and its actors) and a conflict mapping, which is more analytical (i.e. it explains why the conflict takes the form, e.g. why there are latent or manifest forms of violence). Hereby, the latter mentioned perspectives constitute an approach for carrying out conflict analysis on a local level of decision-making and linking it to international and national governance processes.

As both SA and Conflict Analysis are complex and of a heterogeneous nature, given the interrelationships between the different actors and the interacting factors influencing water governance and conflict patterns, there is no mechanistic way of applying a Stakeholder or Conflict Analysis methodology. However, the particular selected methods reflect the specific focus of the chosen research project. The overall circumstances which are caused by the filed research are portrayed within the overall chosen methodological framework. These include, inter alia, the resources available to the researcher, their respective skills and the quality to access them during the research. As mentioned earlier, the specific circumstances of the research project, particularly the planning and conduction of field work along with the quality to access information and to deal with challenges, are discussed in further detail.

1.3.2 Case Selection

Overall, the case selection was based on the theoretical model and the conducted literature review on natural resource conflicts with a particular focus on water and on further information derived from the study of secondary literature. In a first step, the aim was to compile an overview of countries in sub-Saharan Africa and then later in the East African Community. During that step, five general criteria for case selection have been identified and a list of possible water bodies have been prepared: (I) water bodies need to be internal;[3] (II) the water body as well as the water body's surroundings are almost homogenous;[4] (III) national and international political and economic actors are involved in on-the-ground product processing and there has been, therefore, a change in the environmental situation as well as population composition; (IV) occurrence of low-level (sometimes violent) inner-state conflicts since the turn of the millennium (ACLED 2021); and (V) water bodies must be accessible in terms of location and safety.

Accordingly, the first step was to compile an overview of countries related to water shortages and local-level conflicts within Sub-Saharan Africa. This overview served as a basis for the selection of a sub-region (East Africa) and a selection of countries (Kenya and Uganda) and its water bodies, respectively. In a second step, interviews and observations have been conducted during a five-week Scoping Trip to Kenya and Uganda in July and August 2018. The data retrieved from the interviews and observations of the lake sites served as a source of information regarding the potential of the previously chosen lakes and, therewith, more in-depth analysis and to concretise further research. A precise description of the planning and execution of the field work is discussed in further detail. In a third step, another in-depth literature review of studies from international organizations and donors combined with a review of the academic literature on natural resource governance, water shortages and inner-state conflicts in Kenya and Uganda was conducted. Finally, the previous steps have been combined and Lake Naivasha in Kenya and Lake Wamala in Uganda have been chosen as the case studies. Both research sites conformed in terms of the interaction between international and national actors' engagement of outsourcing and commercializing the lakes' resources. The purpose of this outsourcing is to increase the economic development of the region. Furthermore, both case studies matched in terms of local actors playing a key role in the eruption and continuation of local-level conflicts.

Within the wider water shortages-conflict framework, the two cases studied in this book – Lake Naivasha in Kenya and Lake Wamala in Uganda – can be considered to constitute typical or representative cases as they form part of a larger group of (background)[5] cases illustrating normal or average features (Shakir 2002). In this study, background cases are integrated into the analysis to detect essential characteristics of the water shortage-conflict nexus. These background cases include information from Karamoja and Fort Portal in Uganda as well as Turkana Lake and the wider Rift Valley in Kenya.

Two selected cases form part of a larger multiple-case study research designs. According to Yin (2003), the main objective is to use the selected cases to develop or predict a larger study design based on similar results (literal replication) using a small sample of cases (Yin 2003). Instead of conducting a 'rigorous, systematic comparative analysis' (Lund 2014: 226), the purpose of a two-case study is to enhance the analytical coherence of a two-case study design, compared to one single-case design. Yin furthermore argues that the contexts of the case examples can differ (Yin 2003), which is also partly the case in this book. However, if the researcher arrives at common conclusions from both cases despite the different framework conditions, then the external generalizability of the results can be expanded. Hence, the adopted two-case study design in this book serves to strengthen the conclusions obtained by the empirical analysis.

The primary data used during the research process has been obtained through semi-structured and structured interviews with different groups of actors, from experts to representatives of national and sub-national political and economic organizations to local resource users around Lake Naivasha and Lake Wamala. The main group of respondents were local resource users (i.e. farmers, fishermen, pastoralists and villagers), along with information from other actors working within the national water industry, as well as over-sea representatives interested in not only water but also other natural resources in general. The data collected from within the natural resources and water sector was supplemented or triangulated with sources of information from outside the water industry and included, for example, international organizations, national political and economic companies, public authorities or non-governmental organizations (NGOs). Other crucial data sources covered not only interviews with flower farm, company representatives and researchers but also 'informal interviews' with consultants and development organizations during coffee or dinner appointments. Lastly, direct observations of the lake sites as well as factory visits were another valuable source of data collection during the different fieldworks. The water sector literature, especially the water governance acts, has been particularly important to the research in terms of keeping abreast with developments of the water sector and the advanced devolution in both countries. The latter has been of importance, given the highly dynamic nature of the water sector, shifting political responsibilities and its potential for economic prospects. Therewith, political competencies based on the devolution concept have been changed rapidly over the course of time, leading also to conflicts between national, county/district and sub-county/sub-district responsibilities.

1.3.3 Analysis of Data

The analytical strategy adopted in this book, that is, defining priorities for what to analyse and why, as well as what to do with the data (Yin 2003), was guided by the SA and Conflict Analysis framework. Referring to the first question of what to analyse,[6] this was addressed through the questions in the interview guides, which are centred around the key dimensions of stakeholder and conflict analysis, as mentioned earlier. The second set of questions aimed to determine not only the stakeholders' interests with regard to water use but also the influence on water legislation and its availability. For the empirical analysis, the identified stakeholders were grouped by category (see Figures 3.3 and 3.4, Chapter 3). The criteria for selecting the stakeholders were as follows: (I) the daily use of water for private or commercial purposes and (II) the influence of how water is distributed and made accessible in the region.

The second issue, what to do with and how to analyse the data, involved several steps or phases based on the progress of work. During the fieldwork, most interviews were recorded using a Dictaphone. Due to language barriers, which are discussed in the following section, not all interviews were audio-recorded because the author considered that it would limit and reduce the quality of the information conveyed. The interviews during the group discussions and during the visits of the landing beaches and lake sites were recorded in note form and typed up, as well as reworked subsequently by the researcher herself. Further, it was sometimes difficult to make sense of the state's involvement in private businesses through joint ventures, cartridge networks or the involvement of officials and civil servants in fishing, flower farming and tourism.[7] Consequently, the researcher decided to supplement the findings from within official 'external' sources of information, including industry experts, international organizations, development organizations and NGOs.

The data collection during fieldwork was subjected to a more in-depth analysis after returning from the field. To transcribe the audio-recorded interviews, the software MAXQDA was used. In the first phase after the transcriptions, the author examined the data by going through all interviews as well as detailed field notes of the group discussions and lake site observations to tease out salient themes and issues. In the second phase, the texts were categorized with codes and sub-codes relating to the different groups of questions in the interview guidelines (e.g. conflict dynamics and forms of violence, aspects of the water sector, level of influence and interest of the different stakeholders). The separate analysis of the data retrieved from the Scoping Trip and the Three-Month Fieldwork served as a cross-check of the data received and the conclusions drawn. It further helped to 'test' whether the findings resonate with those of other studies dealing with the water shortage-conflict nexus. In a last phase, as part of writing up the final book, all data was again subjected to a final scrutiny in order to capture details and characteristics that have not been included in the secondary literature.

In addition to the primary data obtained from the interviews, this work is based on the evaluation of quantitative data. The statistical analysis of existing climate, water and conflict data is used as the main source for data analysis. As previously mentioned, this book deals in particular with low- and sub-national conflict levels, and therefore, conflict data from ACLED (Armed Conflict Location and Event Data Project) as well as from SCAD (Social Conflict Analysis Database) is considered[8] (ACLED 2021; SCAD 2019). Initially, it was intended for the data to be received directly from the Meteorological Departments of Kenya and Uganda. As the negotiations have failed, data from the Climate Research Unit in East Anglia and the Climate Change Program of the World Bank Group was used (East Anglia 2019; World Bank Group Climate Change 2019). Hereby, the calculations of the statistical quantities median, arithmetic mean, variance and standard deviation are used to make a statement about the development of temperature and precipitation and the water levels in the past 50 years as well as to make a prognosis for the coming years. Furthermore, the representation of these calculations in diagrams allows for a better illustration of the mentioned data. Quantitative data analysis is primarily used to answer the question whether there is a link between water shortages related to climate change and conflict on the one hand and the temporal and local resolution of conflicts on the other hand.

1.3.4 Reflections on Fieldwork in Kenya and Uganda

Often, fieldwork is said to be a linear and rational process, originating in a research design as well as implemented through data collection and resulting finally in published work. However, as Sæther (2006) argued, the actual experience of field research is often far from these depictions (Sæther 2006). One of the most important features of doing field research in African countries, such as Kenya and Uganda,[9] is the researcher's dependence on local partners and assistants. Research assistants work as translators and gatekeepers because they not only know the research areas but are also respected by the communities. Furthermore, they can reduce the local people's suspicions about the researcher's intentions (Stevano and Deane 2017). As previously mentioned, working through local academic partners as well as local organizations and individuals to access communities and information channels is a prerequisite when conducting field research. However, this relationship is rarely highlighted as the difficulties might undermine the validity of the research in general. Thørgersen and Heimer (2006) mention that this tendency might blur the picture of how fieldwork was actually conducted and conceal how the foreign researcher is dependent on local assistance (Thørgersen and Heimer 2006).

Collaborating with local partners and research assistants has both advantages and disadvantages. On the one hand, it provides grounds for mutual benefits and opportunities including shared learning, the exchange of ideas, research partnerships for both sides and a privileged access to interviews and other sources of information for the foreign researcher (Scott 2006). On the other hand, working through these official channels also restricts the autonomy of the researcher. Working under the mentioned circumstances has led researchers in the past to resort to variants of what Thørgersen and Heimer call guerrilla interviewing. Hereby, they refer to unsupervised and spontaneous yet structured interviews, as well as participant observation which present themselves to the research, as ad hoc opportunities (Thørgersen and Heimer 2006).

Although it would be an exaggeration to classify the field research as guerrilla interviewing, the research was constrained and at the same time made possible by the realities of working within the framework of official channels via local partners and institutions. The latter influenced the selection of respondents during the actual stay and concerned, therefore primarily the selection of local resource users. Even though some of the interviews were spontaneously arranged at a lake site or after a concluded interview, which is why a structured interview was not always possible to conduct, this method of data collection allowed the researcher of this book to conduct independent interviews and observations. These ad hoc arranged interviews enabled the respondents to discuss social and political problems more openly. In addition, and as previously mentioned, alternative sources of information to these channels were used to strengthen the reliability of the research data.

To sum up, the methodological approach used to conduct research is based on a two-case study methodology within the broader methodological framework of qualitative and interpretive social research. The importance of primary data collection through fieldwork highlights not only the role of political sensitivity by accessing information through official channels and language barriers but also the importance of supplementing the data with other sources of information. However, most interviews were conducted in English in order to strengthen the reliability of primary data. By applying different strategies, including the use of other information channels, sharpening the observations reflecting upon and being transparent about the research process and its challenges, it is possible to increase the quality and reliability of the research.

1.4 Book Outline

This book consists of six chapters. Thematically, the book can be divided into two parts. Methodology and theoretical concepts (Chapter 1 to Chapter 2) and the empirical analysis (Chapters 3 to 6). Each chapter builds on the prior one(s). Additionally, each empirical chapter pursues a specific objective and, therefore, can also be read independently. In Chapter 2, the key theoretical concept is introduced. The first section introduces the SA approach. The second section discusses the main aspects of conflict analysis, conflict assessment and conflict mapping. The chapter concludes with a brief summary of the theoretical frameworks which are used to examine the research objectives studied in this book.

Chapter 3 is divided into two overall sections. The first section 'maps' the hydrology of Kenya and Uganda, starting with a delineation of the current water status, including its occurrence and distribution. This is followed by a description of the projected rainfall and temperature developments and possible implications of water availability and accessibility. The second section analyses the research areas and their respective stakeholders. The SA framework provides the basis for the discussion of the stakeholder's interaction and interests in the lake's resources. The chapter is divided into two country sections that analyse the research sites and their stakeholders in relation to the same parameters. Each country section starts with an account of the broader geographical and climatic context followed by an overview of the main stakeholder groups and their key interests in the water resource and their influence to integrate water-related policies into the overall management of the lake. More specifically, it traces how the institutional framework and the power of stakeholders in key export and economic markets and domestic regulatory processes influence both water access and water distribution and their geographical position in the lake basins.

In Chapter 4, the vulnerability framework is taken as a basis to understand why conflicts over water erupt between the local stakeholders. The chapter first looks into the climatic conditions for both Lake Naivasha and Lake Wamala. Thereafter, the chapter analyses the internal economic and governance dynamics of the three respective aspects of social, contextual and governmental. Thereby, it identifies the levels of adaptive capacity and sensitivity of the determined stakeholder groups in the previous chapter. Based on the discussion, a first model framework is drafted, which might support the subsequent analysis of the pastoralist–farmer and fishermen–farmer conflicts. The final part of the analysis reflects on the role of the broader institutional, regulatory, economic and social system in shaping the overall conflict risk within and across the stakeholder groups. The discussion will show that the risk over the outbreak of water-related conflict is the highest within the local stakeholder group.

Chapter 5 addresses various aspects of the pastoralist and small-holder farmer conflict at Lake Naivasha and the fishermen and small-holder farmer conflict at Lake Wamala. The conflicts and their dynamics are closely related and influenced by the institutional framework, governance dynamics and the economic agendas described in Chapter 5. Based on the analysis of conflict data and supplemented with qualitative research, a hypothesis is developed, which explains the occurrence of non-violent and violent conflicts in water abundant areas which are, furthermore, independent from the climatic conditions. It highlights that the negative effects of the conflicts reach an extent which undermines the adaptive capacity of the local stakeholders and ultimately accelerates violent behaviour.

Chapter 6 draws final conclusions, starting with a discussion of the main drivers of water-related conflicts. Based on the empirical findings of the research, the second section looks into how the institutional framework and social and economic drivers contextualize and drive water-related conflicts. A final model framework is presented which combines the results of the empirical chapters and links the identified stakeholders to the factors which drive vulnerabilities and, therewith, ultimately lead to the eruption of water-related conflicts. Based on the model framework and the hypothesis developed, policy recommendations are provided to mitigate water-related conflicts.

Notes

1 Ethiopia, Rwanda, South Sudan, Somalia and Tanzania are also part of East Africa. Tanzania, Somalia and South Sudan recorded more severe declines in permanent water bodies compared to Kenya and Uganda. The state of Ethiopia's and Rwanda's water bodies are similar to Kenya and Uganda; however, in terms of accessibility and security aspects, Ethiopia and Rwanda have not been chosen.

2 See Chapter 2 for a detailed discussion of the case selection. See Chapters 2 and 6 for a detailed description of the research area.
3 There is evidence that water shortages at inter-state water bodies provide an impulse for co-operation as riparian states share a complex network of environmental, economic, political and security interdependencies (see, e.g. Carius et al. 2004; Dinar et al. 2011; Almer et al. 2017; Transboundary Freshwater Dispute Database 2009). Studying the water shortages-conflict nexus at inner-state water bodies therefore has the potential to be particularly informative.
4 The term 'homogenous' implies that (a) the inhabitants of the lake's surroundings are mainly pastoralists, farmers, villagers and fishermen and (b) the water levels of the selected water bodies show similar changes of their water levels over a longer period of time.
5 Background cases are defined as not being cases per se, instead, they are integrated into the analysis often in an informal manner but play an important role in case study methodology (Seawright and Gerring 2008).
6 The aim of the study was to understand the linkage and interplay between the institutional framework, the actor's vulnerabilities to water shortages and, therewith, the eruption of non-violent and violent water-related conflicts.
7 The duality of interests often pervades the industry and other economic affairs as political officials are often mandated to oversee the working procedures of the businesses but also having political and economic interests due to their personal involvement in processing, exporting or political affairs (Stonich and Bailey 2000).
8 The author of this book is aware of other conflict datasets, governmental (e.g. CEWARN: Conflict Early Warning and Response Mechanism) or non-governmental conflict datasets (e.g. UCDP: Uppsala Conflict Data Program, Resource Conflict Institute) are not taken into consideration given their focus on conflicts with at least 25 battle-related deaths per year as well as the involvement of at least one governmental actor.
9 Although Uganda was only a British protectorate and never possessed the status of a British colony like Kenya, nevertheless influences and structures of the British were established in the country.

Bibliography

Ali, A. M., Shafiee, M. E. and Zechmann Berglund, E. (2017): Agent-based modeling to simulate the dynamics of urban water supply: Climate, population growth, and water shortages. *Sustainable Cities and Society* 28, 420–434.

Almer, C., Laurent-Lucchettie, J. and Oechslin, M. (2017): Water scarcity and rioting: Disaggregated evidence from Sub-Saharan Africa. *Journal of Environmental Economics and Management* 86, 193–209.

Armed Conflict Location and Event Dataset (2021): *Armed Conflict Location & Event Project. Bringing Clarity to Crisis*. At: www.acleddata.com/ [retrieved last 10 May 2021].

Busby, J. W., Smith, T. G., White, K. L. and Strange, S. M. (2012): Locating climate insecurity: Where are the most vulnerable places in Africa? In: Scheffran, J., Brzoska, M., Brauch, H. G., Link, P. M. and Schilling, J. (Eds.): *Climate Change, Human Security and Violent Conflict*. Springer. Berlin, Heidelberg.

Carius, A., Dabelko, G. and Wolf, A. (2004): Water, Conflict, and Cooperation. ECSP Report, Policy Brief 10.

Dinar, S., Dinar, A. and Kurukulasuriya, P. (2011): Scarcity and cooperation along international rivers: An empirical assessment of bilateral agreements. *International Studies Quarterly* 5 (3), 809–833.

Döring, N. and Bortz, J. (2016): *Forschungsmethoden und Evaluation in den Sozial- und Humanwissenschaften*. Fifth Edition. Springer. Heidelberg.

European Commission (2020): *Global Surface Water Explorer*. At: https://global-surface-water.appspot.com/ [retrieved last 8 May 2020].

Flyvbjerg, B. (2001): *Making Social Science Matter: Why Social Inquiry Fails and How It Can Succeed Again*. Cambridge University Press. Cambridge.

Flyvbjerg, B. (2006): Five misunderstandings about case-study research. *Qualitative Inquiry* 12 (2), 219–245.

Food and Agriculture Organization of the United Nations (2020): *Aquastat – FAO's Global Information System on Water and Agriculture*. At: www.fao.org/aquastat/en/ [retrieved last 8 May 2020].

Gabzdylova, B., Raffensperger, J. F. and Castka, P. (2009): Sustainability in the New Zealand wine industry: Drivers, stakeholders and practices. *Journal of Cleaner Production* 17 (11), 992–998.

Gerring, J. (2007): *Case Study Research. Principles and Practices*. Cambridge University Press. Cambridge.

Gerring, J. (2010): Case selection for case-study analysis: Qualitative and quantitative techniques. In: Box-Steffensmeier, J. M., Brady, H. E. and Collier, D. (Eds.): *The Oxford Handbook of Political Methodology*. Oxford University Press. Oxford.

Gerten, D., Lucht, W., Ostberg, S., Heinke, J., Kowarsch, M., Kreft, H., Kundzewicz, Z. W., Rstgooy, J., Warren, R. and Schellnhuber, H. J. (2013): Asynchronous exposure to global warming: Freshwater resources and terrestrial eco-systems. *Environmental Research Letters* 8 (3), 034032.

Gleditsch, N. P. (2012): Special issue on climate change and conflict. *Journal of Peace Research* 49 (1), 3–9.

Grimble, R. and Wellard, K. (1997): Stakeholder methodologies in natural resource management: A review of principles, contexts, experiences and opportunities. *Agricultural Systems* 55 (2), 173–193.

Hitzler, R. and Honer, A. (1997): *Sozialwissenschaftliche Hermeneutik*. Leske + Budrich. Opladen.

Hsiang, S. M., Burke, M. and Miguel, E. (2013): Quantifying the influence of climate on human conflict. *Science* 341 (6151), 1–14.

International Panel on Climate Change (2007): *Climate Change 2007 – Impacts, Adaptation and Vulnerability. Contribution of Working Group II to the Fourth Assessment Report of the IPCC*. IPCC. Geneva.

International Panel on Climate Change (2014): *Climate Change 2014. Synthesis Report. Summary for Policymakers. Contribution of Working Groups I, II and III to the Fifth Assessment Report of the Intergovernmental Panel on Climate Change*. IPCC. Geneva.

International Panel on Climate Change (2018a): Impacts of 1.5°C global warming on natural and human systems. In: *Global Warming of 1.5°C. An IPCC Special Report on the Impacts of Global Warming of 1.5°C above Pre-industrial Levels and Related Global Greenhouse Gas Emission Pathways, in the Context of Strengthening the Global Response to the Threat of Climate Change, Sustainable Development, and Efforts to Eradicate Poverty*. IPCC. Geneva.

International Panel on Climate Change (2018b): Freshwater resources. In: *Climate Change 2014: Impacts, Adaptation, and Vulnerability. Part A: Global and Sectoral Aspects. Contribution of Working Group II to the Fifth Assessment Report of the Intergovernmental Panel on Climate Change*. Cambridge University Press. Cambridge.

Kleemann, F., Krähnke, U. and Matuschek, I. (2009): *Interpretative Sozialforschung. Eine praxisorientierte Einführung*. VS Verlag für Sozialwissenschaften. Wiesbaden.

Kummu, M., Guillaume, J. H. A., de Moel, H., Eisner, S., Flörke, M., Porkka, M., Siebert, S., Veldkamp, T. I. E. and Ward, P. J. (2016): The world's road to water scarcity: Shortage and stress in the 20th century and pathways towards sustainability. *Scientific Reports* 6 (38495).

Lund, C. (2014): Of what is this a case? Analytical movements in qualitative social science research. *Human Organization* 73 (3), 224–234.

Methmann, C. and Oels, A. (2014): Vulnerability. In: Death, C. (Ed.): *Critical Environmental Politics*. Routledge. London.

O'Connell, E. (2017): Towards adaptation of water resource systems to climatic and socio-economic change. *Water Resources Management* 31, 2965–2984.

Odell, J. S. (2001): Case study methods in international political economy. *International Studies Perspectives* 2, 161–176.

Ogada, J. O., Krhoda, G. O., Van der Veen, A. and Marani, M. (2017): Managing resources through stakeholder networks: Collaborative water governance for Lake Naivasha basin, Kenya. *Water International* 42 (3), 271–290.

Pacific Institute (2020): *Water Conflict Chronology*. At: www.worldwater.org/conflict/map/ [retrieved last 8 May 2020].

Ropers, N. (2008): *Systemic Conflict Transformation: Reflections on the Conflict and Peace Process in Sri Lanka*. Berghof Handbook Dialogue No. 6. Berlin.

Rosenthal, G. (2018): *Interpretive Social Research. An Introduction*. Göttingen University Press. Göttingen.

Sæther, E. (2006): Fieldwork as coping and learning. In: Heimer, M. and Thøgersen, S. (Eds.): *Doing Fieldwork in China*. NIAS Press. Copenhagen.

Schewe, J., Heinke, J., Gerten, D., Haddeland, I., Arnell, N. W., Clark, D. B., Dankers, R., Eisner, S., Fekete, B. M., Colón-Gonzáles, F. J., Gosling, S. N., Kin, H., Liu, X., Masaki, Y., Portmann, F. T., Satoh, Y., Stacke, T., Tang, Q., Wada, Y., Wisser, D., Albrecht, T., Frieler, K., Piontek, F., Warszawski, L. and Kabat, P. (2014): Multimodel assessment of water scarcity under climate change. *Proceedings of the National Academy of Sciences* 111 (9), 3245–3250.

Scott, S., Miller, F. and Lloyd, K. (2006): Doing fieldwork in development geography: Research culture and research spaces in Vietnam. *Geographical Research* 44 (1), 28–40.

Seawright, J. and Gerring, J. (2008): Case selection techniques in case study research: A menu of qualitative and quantitative options. *Political Research Quarterly* 61 (2), 294–308.

Shakir, M. (2002): The selection of case studies: Strategies and their application to IS implementation case studies. *Research Letters in the Information and Mathematical Sciences* 3, 191–198.

Social Conflict Analysis Database (2019): *Social Conflict Analysis Database*. At: www.strausscenter.org/scad.html [retrieved last 10 May 2020].

Stevano, S. and Deane, K. (2017): The role of research assistants in qualitative and cross-cultural social science research. In: Liamputtong, P. (Ed.): *Handbook of Re-search Methods in Health Social Sciences*. Springer. Singapore.

Stonich, S. C. and Bailey, C. (2000): Resisting the blue revolution: Contending coalitions surrounding industrial shrimp farming. *Human Organization* 59 (1), 23–36.

Strauss, A. L. and Corbin, J. M. (1990): *Basics of Qualitative Research*. Sage. Newbury Park.

Sumila, T. C. A., Pires, G. F., Fontes, V. C. and Costa, M. H. (2017): Sources of water vapor to economically relevant regions in Amazonia and the effect of deforestation. *Journal of Hydrometorology* 18 (6).

Thøgersen, S. and Heimer, M. (2006): Introduction. In: Heimer, M. and Thøgersen, S. (Eds.): *Doing Fieldwork in China*. NIAS Press. Copenhagen.

Transboundary Freshwater Dispute Database (2019): *International Freshwater Treaties*. At: http://gis.nacse.org/tfdd/treaties.php [retrieved last 29 January 2019].

Transparency International (2020): *Corruption Perception Index 2019*. At: www.transparency.org/country/KEN [retrieved last 19 August 2020].

United Nations (2020): *United Nations Decade for Deserts and the Fight Against Desertification*. At: www.un.org/en/events/desertification_decade/whynow.shtml [retrieved last 8 May 2020].

United Nations Department Economic and Social Affairs (2014): *International Decade for Action 'Water for Life' 2005–2015*. At: www.un.org/waterforlifedecade/scarcity.shtml [retrieved last 8 May 2020].

United Nations Department Economic and Social Affairs (2020): *Population Dynamics. World Population Prospects*. At: https://population.un.org/wpp/Graphs/Probabilistic/POP/TOT/947 [retrieved last 8 May 2020].

United Nations Framework for Climate Change (2011): *Climate Change and Freshwater Resources. A Synthesis of Adaptation Actions Undertaken by Nairobi Work Programme Partner Organizations*. Nairobi. United Nations.

United Nations Framework for Climate Change (2015): *The Paris Agreement*. United Nations. New York.

University of East Anglia (2019): *Climatic Research Unit*. At: www.cru.uea.ac.uk/data [retrieved last 10 May 2020].

Vergopolan, N. and Fisher, J. B. (2016): The impact of deforestation on the hydrological cycle in Amazonia as observed from remote sensing. *International Journal of Remote Sensing* 37 (22), 5412–5430.

Wilson, T. P. (1970): Normative and interpretative paradigms in sociology. In: Douglas, J. D. (Ed.): *Understanding Everyday Life. Toward the Reconstruction of Sociological Knowledge*. Aldine. Chicago.

World Bank Group (2019): *Climate Change Knowledge Portal for Development Practitioners and Policy Makers*. At: https://climateknowledgeportal.worldbank.org/ [retrieved last 10 May 2020].

World Bank Group (2020): *Agriculture, Forestry, and Fishing, Value Added (% of GDP) – Uganda*. At: https://data.worldbank.org/indicator/NV.AGR.TOTL.ZS?locations=UG [retrieved last 8 February 2020].

Yin, R. K. (2003): *Case Study Research: Design and Methods*. Third Edition. Sage Publications. London.

2

METHOD AND FRAMEWORK

The Tool to Assess the Multi-level Stakeholder Network

This chapter focuses on the conceptual framework applied in this study. Stakeholder and conflict analyses are the key guiding concepts. Since the aim is also to understand the reasons for the outbreak of conflicts especially between local resource users, key theoretical approaches discussing the natural resource and conflict nexus are briefly touched upon. Understanding different actors' vulnerabilities and resilience on natural resource decision making requires the relevant knowledge to assess the compatibility of the interests in natural resources and the influences to transform these interests into influences on decision-making. In the 1980s, Ostrom applied the term 'polycentricity' to outline governance[1] structures that 'take each other into account in [a] competitive relationship, [in which] contractual and cooperative undertakings . . . have . . . central mechanisms to resolve conflicts' (Ostrom et al. 1961: 831). Based on McGinnis and Ostrom, polycentric governance 'requires a complex combination of multiple levels and diverse types of organizations drawn from the public, private, and voluntary sectors that have overlapping realms' (McGinnis and Ostrom 2011: 15). However, in her concept, Ostrom focuses on a micro- to macro-level action building and, thus, does not fully grasp the importance of macro-level actors in the decision-making processes. This is, nonetheless, an important issue to consider when analysing governance structures in developing countries because otherwise, stakeholders who have not been in power to initiate policies in the first place are being blamed for decision-making problems. All the same, Ostrom's concept does not show these power relationships and the connectedness between the different actors.

Accordingly, any decision-making process is not limited to formal governmental bodies anymore. Therefore, additional institutional features beyond the aspects presented in the core concept of Ostrom, as well as key political science theories (e.g. institutionalism or actor-centred institutionalism), need to be integrated. This includes multiple, and often overlapping, actors with different degrees of autonomy

DOI: 10.4324/9781003271291-2

to achieve the core attributes of a certain policy. What follows are acts of co-operation, competition or conflict to achieve a decision-making outcome between the identified overlapping actors. To devise a theoretical insight into the research objectives stated earlier, this chapter lays the foundation for the theoretical considerations of SA.

SA is suitable to analyse stakeholder's social actions and structural characteristics taking into consideration their interests and interactions in a given setting (Ogada et al. 2017). The roots of SA lie within political economy, but nowadays it is also an efficient tool to map natural resource policies and governance processes. The Stakeholder Concept emerged as a prominent theme in the management literature during the 1980s. R. Edward Freeman's book *Strategic Management: A Stakeholder Approach* was a landmark publication bringing attention to a better understanding of stakeholder identification and stakeholder salience within the fields of business ethics and management (Freemann 1984; de Bussy and Kelly 2010). Besides its application to economy and governance processes in general since the 2000s, SA is also seen as an accepted framework to better understand environmental and development problems. Within the field of environmental and development studies, SA defines aspects of social and natural phenomena.

SA is a holistic approach to gain a better understanding of a system characterized by different interests and multiple objectives (Grimble and Wellard 1997). Therefore, it is a useful tool to analyse governance processes developed in response to the challenges which can be derived from the various possibilities of policy formulation and policy analysis. In particular, the stakeholder concept within governmental processes proved to be particularly successful in settings that are characterized by multiple stakeholders, multiple objectives and multiple interests (see Grimble and Wellard 1997; Yang 2014; Ogada et al. 2017). In this book, the term 'stakeholder' is used to describe a group of either unorganized or organized people who 'share a common interest or stake in a particular issue or system' (Grimble and Wellard 1997: 175). Therewith, stakeholders are found at different levels of decision-making, ranging from the global via the national to the regional level. It further covers the concerns of the household or intra-household level as well. More so, both individual stakeholders and groups of any size or aggregation are considered by the approach. Within the governance realm, SA can usefully be applied to complex situations which are characterized by compatibility problems between objectives and the various stakeholders (Ibid. 1997). A recognized method for discussing SA consists of three sub-aspects – to identify the stakeholders, to differentiate between stakeholders, and to investigate the relationship between the stakeholders (Reed et al. 2009). However, before the actors can be identified, the topic and issue in which the stakeholders are involved in have to be defined first (Dougill et al. 2006; Prell et al. 2007).

The theory both identifies different groups of actors affected by environmental change and describes how these actors contribute to environmental change (Reed et al. 2009). Because environmental politics and natural resource governance typically deal with multiple stakeholders and often conflictual interests, SA is a useful framework to discuss the interrelationship between natural resources, water shortages and conflict dynamics. Resource governance, furthermore, can be defined as 'a partnership in which government agencies, local communities and resource users, non-governmental and other stakeholders negotiate, as appropriate to each context, the authority and responsibility for the management of a specific area or set of resources' (Ogada et al. 2017: 271). Based on that definition, SA is an integrated approach to get an understanding of a system, and to assess how the system changes by identifying the key actors or stakeholders and critically evaluate their respective interests in that system regarding a certain topic (Grimble and Wellard 1997). Originally, the term 'interest' was used in an economic sense to measure a gain or loss towards a specific utility or welfare received (Ibid.). However, analysing stakeholders' interests within the setting of natural resource governance, interests are also 'based on action orientation, adhered to by individuals or groups, and they designate the benefits the individual or group can receive from [a] certain object' (Ogada et al. 2017: 278). Because interests cannot be measured directly, the way stakeholders behave and what they do regarding natural resources is used as an indicator to determine their level of interest in the natural resource and the governance process respectively (Ibid.).

As SA can usefully be applied to natural resource governance, it is a matter not only to define the stakeholders and their interests respectively but also to make a fundamental division between those who affect natural resource governance and those who are affected by a decision or action concerning natural resource management (Reed et al. 2009). This essential differentiation into two groups might not be absolute, however, stakeholders can further be categorized according to their relative influence and importance in governance processes. Whereas importance is defined as the needs and interests that are the priorities of aid, influence 'refers to the power certain stakeholders have over the success of a project' (Grimble and Wellard 1997: 176). Similar to interests, the influence of a stakeholder on a governance process cannot be measured directly. Instead, the grade of influence is determined by four indicators: firstly, the role of individuals or groups within the resource governance process; secondly, the amount of resources which are available for the management of natural resources; thirdly, the length of time since one stakeholder or a group of stakeholders has been involved in the management of natural resources; and fourth, the rights given to stakeholders concerning the management of but also the access to natural resources (Ogada et al. 2017). Considering the stakeholders involved regarding their interests, influences and importance, they can be classified into four categories: key players, context settlers, subjects and crowd (see Grimble and Wellard 1997; Eden and Ackermann 1998; De Lopez 2001; figure 2.1).

Key Players have both high interest and high influence and, thus, they play an essential role in natural resource management (Reed et al. 2009; Ogada et al. 2017). Context Settlers are defined as those having low interest despite being highly influential. Because of their importance in governance processes, they cannot be ignored, and hence, they should be monitored and managed (Ibid. 2017).

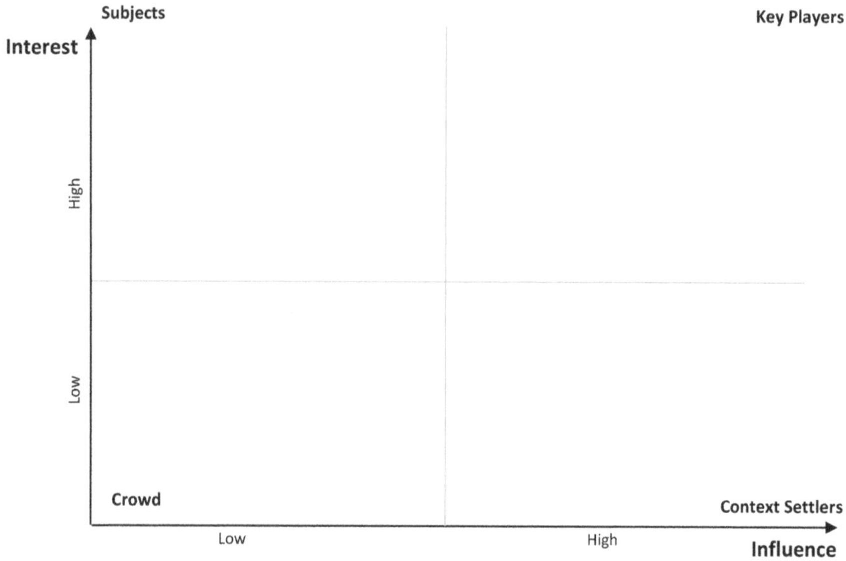

FIGURE 2.1 Stakeholder Matrix

(Source: The Author, based on Grimble and Wellard 1997)

Subjects have high interests but lack influence. Mostly, they are key to every process even though they lack the ability to produce an impact. By definition they are supportive. In order to increase their influence, they should form alliances with other stakeholders (Ibid. 2017). Actors in the Crowd have neither influence nor interest. Therefore, there is little need to consider them in much detail unless they are changing their interest or influence regarding the topic over time. By then, their impact should be taken into account, too (Ibid. 2017).

Due to the focus on natural resource governance and conflicts respectively in this book, the following section teases out the application of SA to the natural resource sector. The natural resource sector is characterized by a complex web of trade-offs, interests and an entanglement of different actors. Resource management situations are cross-cutting systems, given the fact that environmental problems cut across different policy areas, including political, economic, social and administrative units. These different units cover a wide range of stakeholders. Above all, commercial entities and government bodies at regional and national levels can be found throughout economic and political units. Individuals and local community groups are primarily involved within the social level of analysis. Poverty and under-representation are not to be underestimated aspects that influence the classification of stakeholders in the interests and influence matrix. Dealing with environmental problems or the natural resource sector in general, resources such as land, water, pasture and forests are essential to sustain life for the majority of the world's poorest people. SA can further highlight the needs and interests of those people who are directly dependent on natural products. This makes it possible to show to what extent these actors are underrepresented in terms of political and economic power to influence local decision-making processes (Grimble and Wellard 1997). Due to these different sets of actor groups and intersecting units, overlapping but also conflictive interests are likely to become visible.

Thus, stakeholders with strong economic and political relationships are therefore highly influential and placed central in the influence-interest matrix (mostly Key Players overlapping with Context Settlers). Stakeholders in this category tend to have and exercise power, and are able to play a bridging role to other stakeholder groups (Hanneman and Riddle 2005; Ogada et al. 2017). Context Settlers and Key Players are those stakeholder groups who are linked very well to each other and increase each other's importance in the process (Hanneman and Riddle 2005; Ogada et al. 2017). Taking into account the matrix and the resulting stakeholder classifications, SA in natural resource management is able to highlight the trade-offs that have to be made by the stakeholders given their different objectives (Grimble and Wellard 1997).

Due to this complex web of interests as well as trade-offs between the different interacting stakeholders within the natural resource management schemes, conflictive interests and, therewith, conflicts are likely to erupt. *Conflicts* and *trade-offs* are interlinked concepts within the SA framework. However, a conceptual distinction between the two terms has to be made. While trade-offs relate to a single decision-maker or one decision-making group, conflicts always erupt between at least two individuals or stakeholder groups (Grimble and Wellard 1997). Trade-offs are understood as processes to balance conflictive objectives by one particular stakeholder group. A need to balance conflictive objectives emerges once a stakeholder confronts more than one target towards a resource that cannot be achieved at the same time. The final trade-off has to satisfy all the stakeholders and, therefore, a sacrifice or opportunity cost 'in terms of benefits foregone' (Ibid. 1997: 179) has to be applied by all actors involved.

In comparison to trade-offs, conflicts are defined as competitive situations and potential disagreements between at least two but often more actors. Within the natural resource management sector, they usually arise over the use of one or more mostly scarce resources. As mentioned previously, SA discusses the interrelationship between various stakeholders at different levels of decision-making. Therefore, conflict situations are likely to occur at the micro and macro level, as well as between levels. Micro (=local) conflicts erupt between different stakeholders on the ground, for example, farmers or migrant livestock herders. Moreover, local conflicts may also emerge between on-site and external stakeholders, such as settled farmers and pastoralists. Trade-offs at the micro level usually evolve around the question of how resources could be distributed more equitably between the different actors. Macro (=national) conflicts arise between different national stakeholders but can also erupt between national- and international-level actors. A between-level conflict (micro-macro) occurs if the actions of the local stakeholders are contrary to those of macro-level actors. The tensions between micro and macro actors are likely to intensify where the interests of either group may or may not represent the interests of the greater masses (Ibid. 1997).

To conclude, SA provides a theoretical framework to better understand environmental and developmental problems by engaging through a comparative analysis between different stakeholders, their interests and perspectives at various levels of analysis. The interest-influence matrix helps to unpack the different interests and objectives and assists to easier understand the heart of the problems by identifying incompatibilities between the involved stakeholders. Hereby, one is able to analyse how people relate to each other in the natural resource sector by splitting actors according to the roles they play in the natural resource field. Further, they are assigned role-specific rights, responsibilities and profit margins or revenues. Based on this, it is possible to conclude on cooperative or conflictual behaviour (Salam and Noguchi 2006; Reed et al. 2009). However, SA in itself does not try to make any changes. Instead, it only reflects the groups and interests of the wider society. In order to achieve change, stakeholders must have both the interest and the influence to change something (Reed et al. 2009). If a stronger participation of poorly linked stakeholders is going to be achieved or if trade-offs shall be negotiated, this is likely to take place outside the direct realm of SA.

While the vulnerability framework discusses the likelihood of a region or a community to be exposed to environmental changes in general, SA enables for a better differentiation of the actors living within this context. The classification into the four groups of actors allows for a detailed attribution of the different vulnerability types to the individual actors. It further simplifies the analysis of why certain groups of actors are more vulnerable to external influences than others. As it will be shown, the interpretation of conflictual behaviour is closely linked to SA. The relationship between conflict assessment and SA was introduced by Ramirez (1999). He stated that the initial engagement between these two frameworks is either driven by an actor or driven by an intervener. The actor seeks to understand the dynamics and issues of a particular situation before intervening in a non-conflict situation (= SA). The intervener, by contrast, is faced with a social conflict situation (= Conflict Assessment) (Ramirez 1999).

The other key concept applied is conflict analysis. Conflict analysis and mapping are used to discuss the social implications and the risks associated over the eruption of conflicts (Methmann and Oels 2014). Risk is summarized in the equation *risk = vulnerability of the actors x natural hazard* and is commonly understood as the outcome of the relationship between the likelihood of an event to happen and its general impact on the society. A relevant risk in this study refers to the eruption of conflicts between local resource users which erupt over water availability (Scheffran et al. 2014). Conflict for the purpose of this study emerges between at least two actors who fail to achieve their goals and to manage their differences to tolerable levels due to incompatible positions and interests.

In order to make the interactions and interrelationships of the conflict actors in a conflict visible, the drawing of a conflict map is the other essential part of a systemic conflict analysis (Fisher et al. 2000). Conflict mapping is a technique within conflict studies which is done in a similar way to a geographic map (Ropers 2008). The map represents the conflict geographically by placing the parties in relation to both the conflict issue and to the conflict actors. Mapping the conflict dynamics allows also outsiders to the conflict to focus on latent as well as manifest forms of violence. Therewith, one is able to identify potential further outbreaks of violence. Mapping the overall conflict is a method 'of presenting a structured analysis of a particular conflict at a particular moment in time' (Ramsbotham et al. 2011: 89).

Merging Stakeholder and Conflict Analyses, therefore, provides a structured way to (I) identify which stakeholders are involved in a given context and (II) explore how potential inter-actions between these actors affect a certain issue (Poolman et al. 2009). The combination of the frameworks are further a novel tool to describe the eruptions of conflicts as well as to identify how potential interactions between the impact of conflict actors on external stressors might decelerate or accelerate the actors' feasibility to achieve their goals. Thus, very powerful actors are more likely to withstand the impacts of political or socio-economic changes or adverse impacts of other actors 'within the limit of their own capability and sensitivity' (Scheffran et al. 2012: 114).

Note

1 The term 'governance' will be defined in Chapter 5 in more detail.

Bibliography

de Bussy, N. M. and Kelly, L. (2010): Stakeholders, politics and power. *Journal of Communication Management* 14 (4), 289–305.

de Lopez, T. T. (2001): Stakeholder management for conservation projects: A case study of Rem National Park, Cambodia. *Environmental Management* 28, 47–60.

Dougill, A. J., Fraser, E. D. G., Hubacek, K., Holden, J., Fraser, E., Prell, C., Reed, M. S., Stagl, S. and Stringer, L. C. (2006): Learning from doing participatory rural research: Lessons from the Peak District National Park. *Journal of Agricultural Economics* 57 (2), 259–275.

Eden, C. and Ackermann, F. (1998): *Making Strategy: The Journey of Strategic Management*. Sage Publications. London.

Fisher, S., Abdi, D. I., Ludin, J., Smith, R., Williams, S. and Williams, S. (2000): *Working With Conflicts: Skills and Strategies for Action*. Zed Books. London.

Freemann, R. E. (1984): *Strategic Management: A Stakeholder Approach*. Pitman. Boston.

Grimble, R. and Wellard, K. (1997): Stakeholder methodologies in natural resource management: A review of principles, contexts, experiences and opportunities. *Agricultural Systems* 55 (2), 173–193.

Hanneman, R. A. and Riddle, M. (2005): *Introduction to Social Network Methods*. Riverside. At: http://faculty.ucr.edu/~hanneman/ [retrieved last 17 May 2019].

McGinnis, M. D. and Ostrom, E. (2011): Reflections on. Vincent Ostrom, public administration, and polycentricity. *American Political Science Review* 55, 831–842.

Methmann, C. and Oels, A. (2014): Vulnerability. In: Death, C. (Ed.): *Critical Environmental Politics*. Routledge. London.

Ogada, J. O., Krhoda, G. O., Van der Veen, A. and Marani, M. (2017): Managing resources through stakeholder networks: Collaborative water governance for Lake Naivasha basin, Kenya. *Water International* 42 (3), 271–290.

Ostrom, V., Charles, M. T. and Robert, W. (1961): The organization of government in metropolitan areas: A theoretical inquiry. *American Political Science Review* 55, 831–842.

Poolman, M., Munamati, M. and Senzanje, A. (2009): *Stakeholder and Conflict Analysis. Small Reservoirs Toolkit.* At: http://www.smallreservoirs.org/full/toolkit/docs/I%20 02%20Stakeholder%20and%20Conflict%20Analysis_MLA.pdf [retrieved last 2 December 2021].

Prell, C., Hubacek, K., Reed, M. S. and Quinn, C. H. (2007): If you have a hammer everything looks like a nail: 'Traditional' versus participatory model building. *Interdisciplinary Science Reviews* 32 (3), 1–20.

Ramirez, R. (1999): Stakeholder analysis and conflict management. In: Buckles, D. (Ed.): *Cultivating Peace: Conflict and Collaboration in Natural Resource Management.* International Development Research Centre and World Bank. Ottawa.

Ramsbotham, O., Woodhouse, T. and Miall, H. (2011): *Contemporary Conflict Resolution.* Third Edition. Polity. Cambridge.

Reed, M. S., Graves, A. R., Dandy, N., Posthumus, H., Hubacek, K., Morris, J., Prell, C., Quinn, C. H. and Stringer, L. C. (2009): Who's in and why? A typology of stakeholder analysis methods for natural resource management. *Journal of Environmental Management* 90 (5), 1933–1949.

Ropers, N. (2008): *Systemic Conflict Transformation: Reflections on the Conflict and Peace Process in Sri Lanka.* Berghof Hanndbook Dialouge No. 6. Berlin.

Salam, M. A. and Noguchi, T. (2006): Evaluating capacity development for participatory forest management in Bangladesh's Sal forests based on '4Rs' stakeholder analysis. *Forest Policy and Economics* 8 (8), 785–796.

Scheffran, J., Brzoska, M., Brauch, H. G., Link, P. M. and Schilling, J. (2012): *Climate Change, Human Security and Violent Conflict.* Springer. Berlin, Heidelberg.

Scheffran, J., Ide, T. and Schilling, J. (2014): Violent climate or climate of violence? Concepts and relations with focus on Kenya and Sudan. *International Journal of Human Rights* 18 (3), 366–387.

Yang, R. J. (2014): An investigation of stakeholder analysis in urban development projects: Empirical or rationalistic perspectives. *International Journal of Project Management* 32 (5), 838–849.

3

THE GEOGRAPHICAL AND STAKEHOLDER CHARACTERISTICS OF LAKE NAIVASHA AND LAKE WAMALA

3.1 Geographical and Hydrological Characteristics of Lake Naivasha

Lake Naivasha in central Kenya is situated 100 kilometres north-east of Nairobi. It is located in the East African Rift Valley at an altitude of 1,890 metres above the sea level and it is surrounded by Game Parks in the North and two National Parks in the South. It lies at latitude 00 46′ N and longitude 36 22′ E. It is one out of two freshwater lakes in the Rift Valley and is fed by the main input Malewa river which accounts for 80 per cent of the lake's discharge and Gilgil river which accounts for approximately 10 per cent of the total discharge into the lake (Becht et al. 2006). It is a small, shallow freshwater lake and covers a surface area of roughly 160 km^2 (Kundu et al. 2010). The lake consists of three lakes: The main lake is shallow and has a maximum depth of 8 metres. The second part, Crescent Lake, is directly connected to the main lake and is the deepest of the three lakes with a maximum depth of 18 metres. Lake Oloidien is a small lake located in the south-west of the main lake. In the earlier days, it formed part of Lake Naivasha. Nowadays, it is distinct from the lake and receives water only through rainfall. Unlike the other two, Lake Oloidien is not a freshwater lake (Becht et al. 2006).

The climate is humid to sub-humid in the Highlands and semi-arid in the Rift Valley. The mean monthly maximum temperature ranges between 24.6°C and 28.3°C, and a mean monthly minimum temperature between 6.8°C and 8.0°C. The average annual rainfall ranges from about 800 mm in Kinangop plateau to about 300 mm in the rift floor. The region experiences bimodal rainfall patterns. However, its frequency and duration vary due to the effects of climate variability and change considerably, resulting in substantial changes of Lake Naivasha's water levels over the past 50 years (see Figure 3.1; SPEI 2021a). The area was affected by a major drought in 2009. This reduced the lake by one-third of its size. However,

DOI: 10.4324/9781003271291-3

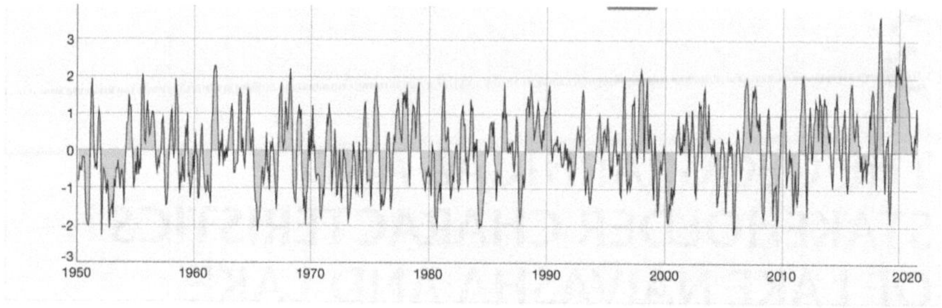

FIGURE 3.1 Rainfall Distribution at Lake Naivasha 1950 to 2021

(*Source:* SPEI 2021a)

in 2018, the lake's water levels were higher than ever before (Ibid. 2021a; European Commission 2020).

Lake Naivasha is located in the Rift Valley Province, in Nakuru County. The lake is furthermore covered by Naivasha District. Historically, Lake Naivasha basin was a Maasai grazing region and the main industry were fishing and farming till 30 years ago. However, after colonization, the basin became part of the British 'white highlands', an area where only British and European settlers were allowed to settle and to own land. After independence, former white settlers and their descendants received land use rights at the northern shorelines for 99 years through a passed law. This law was renewed at the beginning of 2010 allowing them another 99 years of residency. Similarly, members of the Kikuyu tribe were settled around the lake after 1963. Kikuyus were mainly located along the western and south-eastern parts of the lake as those receive the highest amounts of rainfall and therefore, these parts are most suitable for rain-fed agriculture. By that time, roughly 90 per cent of Kenya's economy depended on rain fed agriculture. Nowadays, rain-fed agriculture accounts stand at 75 per cent (Becht et al. 2006). The lake is of international importance, having been declared a Ramsar site in 1995, and is thus protected under the International Convention on Wetlands (Ramsar 2011). Nowadays, it is a significant economic resource due to its social and economic importance, including horticultural production, geothermal power generation and tourism (Becht et al. 2006; Van Oel et al. 2013; Willy et al. 2014; Mulatu et al. 2015).

3.2 Stakeholder Composition and Interests

As already mentioned, Lake Naivasha is at a short distance from Nairobi and, therefore, close to Kenya's major international airport. This geographical proximity came to be appreciated in colonial times when British settlers used the lake and its surroundings as a tourist destination as well as retirement areas. After independence, the national government promoted Lake Naivasha's picturesque setting and advertised the lake's aquatic bird diversity as well as wildlife density and, therefore, the lake also became popular for residents of Nairobi as a weekend escape (Ogada et al. 2017). Furthermore, the national political elite made use of the permeable and fertile soils found around the lake's shorelines and sold or rented land to entrepreneurs and international commercial flower companies. Since the late 1980s, national and international business companies moved in which mainly cultivated flowers for export. Later, at the beginning of the new century, other economic companies rented or bought land to set up geothermal and hydropower companies (Harper et al. 1990, 2002; Harper et al. 2011). The horticulture developments, tourism and industrial activities caused a tremendous shift in the landownership and population composition around the lake. Since the early 2000s, these factors resulted in a sharp population increase in Naivasha town and its surroundings. The promise of labour opportunities has drawn many unskilled workers to the horticulture farms, the hotel industry and the power generation companies. At the beginning of the new century, the population of Naivasha and its surroundings stood at around 300,000 (UNESCO 1999). However, according to recent estimates, the area inhabits between 1 million and 1.2 million people (Kimeta 2019; Peter 2019; Republic of Kenya 2019).

According to Carolina (2002), Lake Naivasha basin is nowadays of high economic and political importance. The basin presents a wide variety of economic activities which are based on the water sources. Additionally, there are many different stakeholders with competing interests in the resources (Carolina 2002). The following part moves on to describe in greater detail the different stakeholders engaged in political, economic and social activities within Naivasha basin. Using SA, this part identifies and classifies the stakeholders as well as their interests in the water and land resources and their power to influence natural resource policies and governance processes (see Table 3.1). The SA will show that international, national, county and local political and economic stakeholders are, above all, involved in water use and governance respectively (see Table 3.1).

Altogether, the SA identified 45 stakeholders from public, private and civil society actors. The selected actors were categorized into nine groups (see Table 3.1). The political actors were divided into the separate groups of agencies and ministries. Even though both groups represent government units, the analysis of the water governance structures revealed different mandates and foci of the institutions working within resource management. The governmental ministries and agencies have the highest number of stakeholders given the inclusion of vertical multi-level interactions at the national, sub-national and local levels. Another notable category is international and national economic actors. Over the past 20 years, economic actors have contributed to ecological changes, thereby causing a shift in the landownership and population around the lake (Becht et al. 2006).

National political actors can be divided into two subgroups. National governmental agencies are interested in not only managing the water resources in a way such that each stakeholder has access to sufficient water but also formulating policies for a better water development on the other hand. More so, they are interested and concerned about improving the urban and rural water provision. They are,

TABLE 3.1 Stakeholder Classification Lake Naivasha

Political Actors (National)

Governmental Agencies

1. National Environmental Management Authority
2. Water Resource Authority
3. Water Tribunal
4. National Water Storage Authority

Governmental Ministries and Departments

5. Government of Kenya
6. Ministry of Water and Sanitation
7. Department for Defense
8. The National Treasury
9. Ministry of Energy
10. Ministry of Devolution and the Arid and Semi-arid Lands
11. Ministry of Lands and Physical Planning
12. Ministry of Agriculture, Livestock, Fishery and Irrigation

Political Actors (Sub-nation/County-Level)

Governmental Agencies

13. Basin Water Resource Committee
14. Water Harvesting and Storage Authority
15. Water Storage Board
16. Lake Naivasha Resource Authority
17. Water Sector Trust Fund
18. Naivasha Water and Sewage Cooperation
19. Water Services Regulatory Board
20. The National Water Conservation and Pipeline Corporation
21. Regional Water Services Boards
22. Water Works Development Agencies
23. Water Basin Committee

Governmental Ministries and Departments

24. County Government Nakuru
25. Ministry for Environment of Nakuru County
26. Department for Water within the County Ministry for Environment

Economic Actors (International)

27. International Flower Farms (Dutch and British)
28. International Investors (e.g. Chinese and Japanese)

Economic Actors (National)

29. Hotels
30. Geothermal industries
31. Power Plants
32. KenGen

Resource Users (Local)

33. Pastoralist Groups
34. Fishing Community
35. Farming Community
36. Villagers at Lake Naivasha and in the upper catchments

NGOs and community-based organizations (Local)

37. Lake Naivasha Riparian Association
38. Lake Naivasha Resource Users Association
39. Community Forest Association
40. Imarisha
41. Lake Naivasha Flower Council
42. Kenyan Wildlife Service
43. Beach Management Unit

I(N)GOs/Agencies (International)

44. World Wildlife Fund
45. World Bank

(*Source*: The Author 2019)

furthermore, interested in the distribution of water to different public and private areas as well as sewage management (Bakari 2018; Mbaisi 2018). The governmental ministries' and departments' actions are driven foremost by economic interests. Since the beginning of the 2000s, their main focus is on the economic development of the area by promoting the tourism sector as well as the export of flowers to increase the Kenyan gross domestic product (GDP). Respondents reported that the government is not directly interested in the land and water resources. Instead, they emphasized that the government, through its ministries, leases the land directly to national and international investors which is located at the shorelines. The leasing agreements last for 99 years. However, the investors have the option to extend the agreements (Ran 2018; Schäfer 2018; Peter 2019). The Land and Water Acts in place grant the government to right to held land and water resources 'in trust for the people of Kenya' (Water Act 2016, Article 5). Therewith, the government, as the owner of the land, is allowed to convert public land into private land if they consider it to be in the best interest of and for its population. Thus, the land is used to increase economic investments. The government's interest in the land and water resources is guided foremost by monetary incentives. Those are received through the performance of national and international businesses which take place in Lake Naivasha basin.

During the past decade, the Ministry of Energy increasingly gained interest in the land and water resources around Lake Naivasha. To harness the country's geothermal capabilities, the government and the ministry pushed hard to provide more households in Kenya with electricity (World Bank Group 2018). Nevertheless, in the case of Lake Naivasha, electricity, especially reliable energy, is needed for economic growth. Employees and members of the flower companies mentioned that the geothermal stem warms the greenhouses and generates electricity not only for the companies, but also for the hotels located at the lake. Due to the rapid set-up of flower farms and hotels, the land around the lake became scarce. Therefore, the Ministry of Energy used the land within near-by Hell's Gate National Park to set up two major geothermal companies. Following the population increase around the lake, big efforts are underway to achieve better energy supply coverage. As a result, another geothermal field is already under development north of Hell's Gate. The power plants inside Hell's Gate helped to generate almost 50 per cent of Kenya's electricity in 2015, with an increasing tendency. Furthermore, these plants also generate roughly 35 per cent of the country's hydropower (USAID 2016; World Bank Group 2018). However, the power generation also impacts on the basin's water supply. Abstraction of water for irrigation is taken through pumping directly from the lake, and accounts for 97.7 per cent of the water abstracted from the lake. The Olkaria Power Project abstracts water from this lake for use in drilling wells and other uses in the power station operations, as well as for domestic use in the company offices and the housing estates. The Olkaria Power Project currently has an abstraction permit limit of 8,000 m^3 per day, which equates to an annual limit of 2,920,000 m^3. In 2000, KenGen abstracted an estimated 919,840 m^3 of water, and this rate increased up to 1,300,620 m^3 in 2008. Since then, the abstraction rate has increased to 1,755,839 m^3 (in 2017), to 2,037,734 m^3 (in 2018), to 1,003,385 m^3 (in 2019) and then to 2,385.391 m^3 (in 2020).

Sub-national governmental ministries, departments and agencies are foremost concerned about a better regulation and management of the lake basins' natural resources. The rise in human population and economic undertakings put considerable constraints on the land and water resources. Accordingly, and resulting from devolution, county-level departments and ministries developed integrated management plans for Lake Naivasha (Harper et al. 2011). The plans emerged from the interests of the stakeholders around the lake regarding the resources. And so, sub-national political stakeholders are interested in improved conservation and sustainable usage of the resource to limit the environmental vulnerability of the area. County and sub-county officers are also interested in an economic development of the area. Economic growth attracts international and national businesses, strengthens locally produced goods, services and derives income. Employees from sub-national agencies report that their interests primarily concern the sustainable development and management of natural resources.

International and national economic actors are interested in a preferential access to the lake's land and water resources to pursue their economic businesses. The wetland associated with the lake provides a habitat for a wide range of aquatic flora and fauna. The water from the lake and the associated aquifers is utilized for, among others, domestic use, watering of livestock, irrigation of agricultural land and geothermal drilling and process activities. When cultivating flowers, roughly 60 m^3 water is necessary to ensure good conditions for flower growing. Reports indicate that the water footprint for one flower is estimated to be 7 to 13 litres (Mekonnen et al. 2012). Fertile agricultural land and sufficient water levels at Lake Naivasha provide ideal conditions for the companies to grow roses cheaply and quickly. The British and Dutch flower companies indicated that outsourcing their industrial sites overseas is, for them, the only way to compete on their home market and to sell high-quality imports (Flower Farm 2019). According to Kuiper and Gemählich (2017), the total area of irrigated land stands at 5,000 ha (Kuiper and Gemählich 2017). However, Lake Naivasha Growers Group (LNGG) estimates the total irrigated area to be much less, probability ranging from 1,500 to 2,000 ha (Kenyan Flower Council 2020).

National companies' interest in the land and water resources is guided by different motives. National hotel owners are keen to set up their hotel sites at the lakes' shorelines to attract international tourists. They advertise 'face-to-face' interactions with the lake's wildlife and ensure the availability of sufficient water in the hotels by using the lake's water resources partly. Geothermal industries are not interested in setting up their sites directly at the lake's shorelines. Tapping geothermal energy is a long and expansive process, therefore finding suitable spots for drilling steam wells is considered top priority. Underground water and high amounts of volcanic rocks are necessary to tap the full potential of geothermal production. The Kenyan rift valley is characterized by extinct volcanic activities (National Geographic 2020) and high groundwater levels. Consequently, the National Parks surrounding Lake Naivasha provide high potential for geothermal businesses. Overall, geothermal companies use water as a fluid to prevent the industrial sites from running dry. It is also partly used for cooling, dust suppression or drilling fluid (Clark et al. 2011).

The group of local resource users can be divided into four groups of actors which determine their interests in the water and land resources. These four groups of local actors are pastoralists, fishermen, small-holder farmers and villagers. Villagers live in Naivasha town and in one of the informal settlements in the lake's closer surroundings. Pastoralists need direct access to the lake and its resources for watering and grazing their cattle. Fishermen are especially interested in free access routes to the lake's shorelines and its landing sites. Small-holder farmers are interested in land which is in close proximity to the lake to pursue their agricultural business and to water the fields. As these three groups of actors also live in Naivasha or the near-by informal settlements, in addition, they need water for everyday activities, for example, cooking, drinking or washing. The informal settlements still lack basic sanitation and water infrastructure owing to the rapid increase in human population. As a consequence, local resource users also rely on the lake's water to cover basic amenities (Community Members Karagita Landing Site 2018).

Generally, international (development) organizations are hardly found at the lake as it is believed that Naivasha is not among the poorest and most conflictual areas in Kenya (Muigai 2019). There is only one major international organization and one international non-governmental institution at Lake Naivasha. The World Wildlife Fund (WWF) is interested in the sustainable management of the resources to decrease the pressures on the lakeshores and, therewith, to limit the effects of already undertaken economic investments, ecological and climate changes (WWF Employee 2018). Employees from the WWF reported that the improvement of the local resource users' livelihood is also in their interest (Muigai 2019) to limit the tensions of conflict as 'they use the resources of the other' (WWF Employee 2018). Over the course of the past three years, the World Bank Group increased their interest in the basin. On the one hand, their interest is in supporting the national government to develop affordable housing and to provide technical assistance (World Bank Group 2019). On the other hand, they focus on the development of the infrastructure network. An employee of a Nairobi-based German development organization indicated that the World Bank Group is interested in funding a 2000 ha large industrial dry cargo site in the south-west of Hell's Gate National Park. Accordingly, their interest in water and land is guided by the improvement of sanitation in the field of development co-operation as well as supporting the national government in managing their capital and ensuring maximum value for money.

National NGOs and other community-based organizations are interested in the use and management of the resources in a way to preserve the environmental stability and to foster the socio-economic development within Lake Naivasha basin. Furthermore, they try to make sure that economic and political infrastructure projects comply with the laws governing the natural resource sector. For example, Lake Naivasha Riparian Association (LNRA) or Lake Naivasha Resource Users

Association (LNRUA) initiates community meetings to raise the local people's awareness about their rights and duties they have in relation to the land and water (Peter 2018). As a consequence of the used chemicals in the flower farms and the inefficiency of Naivasha's sewage companies, LNRUA is keen to mobilize the population to take action against water pollution. Likewise, community-based organizations are interested to mobilize people to demonstrate against the closing of the public corridors leading to the lake. Overall, their interest is in the sensitization about the sustainable use and the conservation of water and land resources (Kimeta 2018; Peter 2018).

3.3 Geographical and Hydrological Characteristics of Lake Wamala

Lake Wamala is located 95 kilometres west of Kampala. It is situated 1,290 metres above the sea level. It lies at latitude 00 19′ N and 31 50′E (Baker et al. 2019). It is among Uganda's freshwater lakes and is fed by four rivers: Nyansi-Kitenga, Kabasuma, Mpamujugu and Bimbye (Kaganga 2018). It is a small shallow lake, covering an area of originally 250 km^2 (Vianny et al. 2016). Lake Wamala is dotted by many islands, which include, among others, Lwanju Island, Mabo Island or Bagwe Island. Lake Wamala has a mean depth of 4.5 metres (Ibid. 2016).

The climate is humid to semi-arid. The mean monthly maximum temperature varies between 22.1°C and 23.1°C. Especially between January and March, the mean monthly minimum temperature ranges between 13°C and 15°C (Baker et al. 2019). The region experiences bimodal rainfall patterns. The average annual rainfall ranges from 1,600 mm to 900 mm (Musinguzi et al. 2015). The IPCC indicates that rainfall patterns are going to increase across East Africa. The SPEI values illustrate that the average rainfall for Lake Wamala are consistent with this general trend (IPCC 2007; see Figure 3.2; SPEI 2021b). However, the severity of drought is progressively increasing within the basin and resulted in tremendous changes of Lake Wamala's water levels. Between 1984 and 1995, the water levels of the lake shrunk to about half its size. Since the turn of the millennium, the

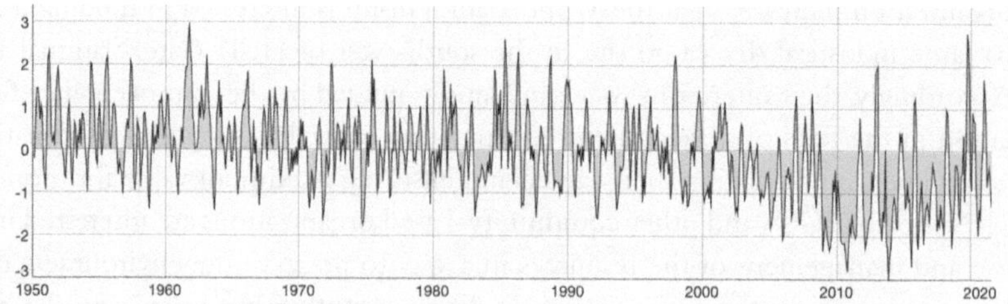

FIGURE 3.2 Rainfall Distribution Lake Wamala 1950 to 2021

(Source: SPEI 2021b)

lake's water levels increased, nevertheless, it did not fully recover to its original size. In 2013, the mean depth stood at only 3.8 metres (UNEP 2009; Natugonza and Ogutu-Ohwayo 2015). Additionally, the severity of drought is more extreme. This assumption is based on the probability of the observed total annual rainfall. Thus, it is assumed that the mean depth of the lake stands nowadays below 3.5 metres, and that the lake covers an area of approximately 180 km^2 (Kitzio 2019).

Lake Wamala is located in the Central Region and is shared by the three districts of Mubende, Gomba and Mityana. Historically, Lake Wamala basin was an artisanal fishing area. In addition to fishing, farming is the main industry for the people residing in one of the four fishing villages, around Lake Wamala or in Mityana town. Before and after independence, the lake was of importance for a number of livelihoods located directly around the lake, directly or indirectly benefitting from the lake's wetland resources and the lake itself (Baker et al. 2019). Compared to other lakes in Uganda, Lake Wamala is relatively small and was, thus, neglected and ignored by national authorities (Kaganga 2019). Over the past 30 years, however, national authorities gained interest in the lake as a result of the growth in human population. Nowadays, the lake is of national importance and a significant economic resource, including agricultural and industrial uses.

3.4 Stakeholder Composition and Interests

Lake Wamala's geographical proximity to Kampala and Uganda's main international airport in Entebbe came to be appreciated by political authorities in a similar way as in the case of Lake Naivasha. In contrast to Lake Naivasha, stakeholders picked up interest in Lake Wamala since the turn of the millennium. After 2000, national and international business companies and NGOs moved in, which carried out deforestation in order to acquire land. Later, national and a few international companies obtained land titles to set up their businesses (Mucunguzi 2019). There are two significant differences between the research objects. Unlike Lake Naivasha, the central government of Uganda did not promote Lake Wamala as a tourist destination (Bbira 2018). Additionally, the central government did not promote the basin as a work destination for unskilled workers in particular. Instead, people residing in one of the small fishing villages have been displaced to create space for economic activities (Twesigye 2019). Therefore, the lake did not experience a sharp population increase as Lake Naivasha did. By contrast, the number of residents at the lake directly decreased. Because of the general population increase and the unavailability of sufficient surface water within a 60 km radius to the lake, the amount of people which depend on the lake's resources, however, increased. It is estimated that 1.5 million people are dependent on the lake's resources currently (Kitzio 2019).

In all, the SA identified 46 actors from public, private and civil society actors (see Table 3.2). The named actors were organized into the same nine groups as in the case of Lake Naivasha. The governmental ministries and agencies have the

TABLE 3.2 Stakeholder Classification Lake Wamala

Political Actors (National)

Governmental Agencies

1. National Environmental Management Authority
2. National Water and Sewage Cooperation
3. Umbrella Organizations for Water Provision
4. Petroleum Authority Uganda
5. Investment Authority

Governmental Ministries and Departments

6. Government of Uganda
7. Ministry of Finance
8. Ministry of Energy and Mineral Development
9. Ministry of Water and Environment
10. Ministry of Land, Housing and Urban Development
11. Ministry of Works and Transport
12. Ministry of Agriculture
13. Parliament
14. Natural Resources Management Group of Parliament

Political Actors (Sub-nation/ County-Level)

Governmental Agencies

Governmental Ministries and Departments

15. Environmental Officer Mityana County
16. Environmental Officer Mubende County
17. Environmental Officer Gomba County (
18. Lake Albert Water Management Zone Officer
19. Forestry District Officer
20. Local District Officer

Economic Actors (International)

21. Coca-Cola
22. Total
23. Chinese Companies/Investors

Economic Actors (National)

24. Ugandan Breweries
25. Sand Miners
26. Gold and Mineral Miners

Resource Users (Local)

27. Villagers/Community
28. Farmers
29. Fishermen
30. Pastoralists

NGOs and community-Based Organizations (Local)

31. Kikandwa Environmental Association
32. Rural Development Media and Communication
33. Uganda Coalition for Sustainable Development
34. National Association of Professional Environmentalists
35. Wetland International
36. Ecological Christian Association
37. Civic Response on Environment and Development

I(N)GOs/Agencies (International)

38. Gesellschaft für Internationale Zusammenarbeit (GIZ)
39. Kreditanstalt für Wiederaufbau (KfW)
40. French Development Agency
41. Italian Agency for Development Cooperation
42. European Union
43. World Bank
44. World Wildlife Fund for Nature
45. International Committee of the Red Cross
46. Oxfam

(*Source*: The Author 2019)

highest number of stakeholders. Similar to Lake Naivasha, the inclusion of vertical multi-level interactions among the national via the sub-national to the local level of decision-making can be cited as the reason for this high number of actors. However, governmental agencies are not represented at the district level.

Politically, Uganda is in the process of decentralizing the natural resource sector. Consequently, the creation of institutional arrangements and water management zones was the top priority (Kitamirike 2019). Another noteworthy difference is the vast amount of international non-governmental and development organizations. The international development organizations support the Ugandan water sector institutions in strengthening 'climate resilience of water supply systems' or promoting 'the institutionalization of catchment-based planning structures' (GIZ 2018).

The interests of the Ugandan governmental departments and ministries are complementary to the Kenyan ones. Consequently, the national authorities' interest in the water and land resources is driven by economic incentives. Villagers and the natural resource officer from Mityana reported that the government's interest is with production for industrial and agricultural uses (Bbira 2018; Kikandwa Community Member 2019). In Kenya, while the horticulture and geothermal industry contributes to the economic development of Lake Naivasha, the Ugandan Ministry of Energy and Mineral Development is interested in sand and gold mining in particular. One representative of an NGO commented that when it comes to the extraction of minerals, areas are prioritized which have more water available than others (Twesigye 2019). Both the Ministry of Energy and Mineral Development and the Ministry of Land, Housing and Urban Development are interested in timber and sand for construction purposes. However, the national ministries and departments are, furthermore, interested in using governmental agencies as intermediaries which delegate land to extractors (Anthony 2019). The Investment Authority and the Petroleum Authority are interested in enabling investors to access the wetlands around Lake Wamala for sand and gold mining, rice growing, underground explorations or as transit-routes for pipelines. Both authorities are keen to secure higher dividends through the undertaken investments in the country (Ibid. 2019).

Sub-national ministries' and departments' interests are akin to the ones of Lake Naivasha's county representatives. The Lake Albert Water Management Zone officer is interested to secure enough funding from the Ministry of Finance to set up the institutional arrangements for the management of the 42 districts that the zone is covering (Kitamirike 2019). Furthermore, the interests of both the catchment officer and the environmental officers from the three districts focus on the improvement of the water quality and the compliance with the law, especially of the local resource users (Bbira 2018; Ibid. 2019). In this regard, the natural resource officer in Mityana stated that the environmental law 'provides a buffer zone of 200 m from the water mark of the lake' to the shoreline (Bbira 2018). Nevertheless, the villagers encroach with their agricultural activities into these buffer zones and contribute to the deterioration of both the quality and quantity of the lake's wetland cover (Ibid. 2018; Kitamirike 2019).

International and national economic actors are interested in pursuing their businesses by using the land and water resources. In particular, international companies are keen to collaborate with national companies. Chinese investors, especially, operate behind the national economic face as it shields them away from the public and private spotting (Moses 2019). Companies like Coca-Cola or Ugandan

Breweries hold two interests. Firstly, the private companies go along with all that brings the return of investment as fast as possible and, therefore, they use resources inconsiderately. Secondly, because these companies are known worldwide, they try to preserve the environment and ensure the marketing of an eco-environmental site through eco-friendly measures (Nakaggwa 2019). Therewith, the aim is to protect the company's name.

Similar to Lake Naivasha, the group of local resource users is split into four groups of actors. Fishermen are foremost interested in staying in one of the three remaining fishing villages which are located around the lake. As already mentioned, fishermen are displaced to areas more distant from the lake as the land is repurposed by the Ministry of Land and the government. Small-holder farmers rely on the water from the lake to water their cropland. Furthermore, villagers and small-holder farmers need the lake's water for everyday activities, including washing, cooking and drinking. Neither the fishing villages nor the communities in the broader environment have access to water infrastructure and, therefore, rely on the lake's water directly or through pumping stations indirectly. Hardly any pastoralists pass Lake Wamala and, therefore, are not considered in much detail (Mutambukah 2019).

In contrast to the stakeholder composition at Lake Naivasha, there are considerably more international (development) organizations around Lake Wamala. Not only German but also Italian and French Development Organizations support the national authorities to enhance water security and sanitation in Uganda. In particular, they are interested to strengthen climate resilience of water supply systems, to develop a sustainable faecal sludge management strategy in both urban and rural areas and to support the training of water supply and sanitation experts (Nakaggwa 2019; Namukose 2019). More so, international development organizations are keen to support the country's water sector institutions to develop Catchment Management Plans. In doing so, they support the deconcentrating process in catchment-based water resource management and promote a better institutionalization of catchment-based planning structures. Likewise, international development organizations develop water resource protection plans to prevent further depletion of the water bodies and to increase their sustainable management (Anders 2018; Namukose 2019). Additionally, the International Committee of the Red Cross (ICRC) is involved in the maintenance of the existing water infrastructure and the set-up of water access points, especially for small-holder farmers and villagers in remote areas (Vergin 2019). The World Bank Group, the European Union (EU) or Development Banks are interested in enabling and supporting state authorities in advancing urban and rural water supply financially (Juenger 2018; Fauvel 2019; Liebig 2020).

National NGOs and community-based organizations are, foremost, interested to promote the conservation of public lands and to raise awareness about the effects of deforestation or unsustainable farming activities on the lake's ecosystem. Their aim is to minimize crop and animal raiding (Olupot 2018). Furthermore, there are incentives to capitalize fruit market products. Other environmental and sustainability organizations are actively engaged to limit the impacts of climate change on agricultural productivity. They are also engaged in contamination control, biodiversity protection and sustainable water use management in the lake's basin (Kimbowa 2018). As a result of increased degradation and deforestation activities, community-based organizations are keen to mobilize local communities to not only take action against water pollution and also hold trainings for and with the communities. A respondent reported that their aim is to train small-holder farmers in sustainable agricultural techniques to avoid further water contamination and siltation or wetland degradations (Ibid. 2018). Other participants indicated that they also try to raise the community's awareness about law enforcement and political responsibilities in the natural resource sector during these meetings (Kamese 2018). Overall, the interviews demonstrated that they are interested in the sensitization of sustainable agricultural activities, the conservation of land and water resources and better law enforcement to protect the basin area and to sustain its biodiversity.

To sum up, the analysis has shown that Kenyan national political actors are foremost interested in an economic development of the area by promoting the tourism sector and the export of flowers. The Ugandan government promotes the exploration of sand and other minerals as well as agricultural products like rice. International and national economic actors are primarily interested in a preferential access to the lake's resources to pursue their economic businesses. Local resource users have different interests regarding the land and water resources around the lakes. On the one hand, pastoralists use land and water for watering and grazing their cattle. Farmers, on the other hand, need land to pursue their agricultural business and to water the fields. Fishermen are interested in free access routes to the Lake's shorelines and villagers need water and land for everyday activities, including cooking, drinking or washing. Sub-national governmental ministries, departments and agencies are foremost concerned about better regulation and the management of the natural resources. Lastly, NGOs and international development organizations hold interest in the sustainability of the basin's resources, the improvement of the local resource users' livelihood and the empowerment of local resource users. Moreover, some community-based organizations are also interested in using and managing the resources in a way to preserve environmental stability as well as to foster the socio-economic development of the area.

The purpose of SA is 'to gain an understanding of multi-stakeholder dynamics in participatory resource management through identification of stakeholders in a particular aspect of the system, and . . . their involvement in decisions on aspects of the system' (Ogada et al. 2017: 275). Consequently, the final part of this chapter focuses on the assessment of the different stakeholders' influences in decision-making processes, thereby classifying the actors into the four different stakeholder categories (see Figures 3.3 and 3.4).

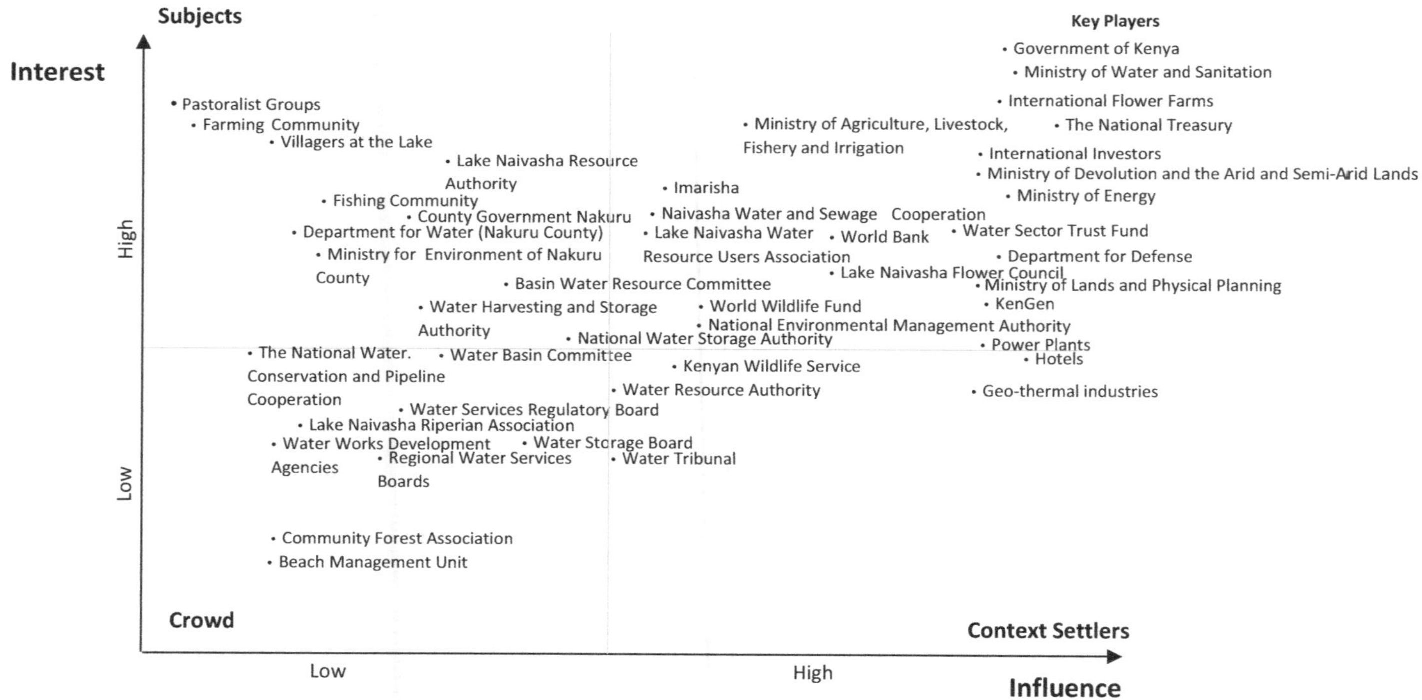

FIGURE 3.3 Influence-Interest Matrix Lake Naivasha

(*Source:* The Author, based on Grimble and Wellard 1997)

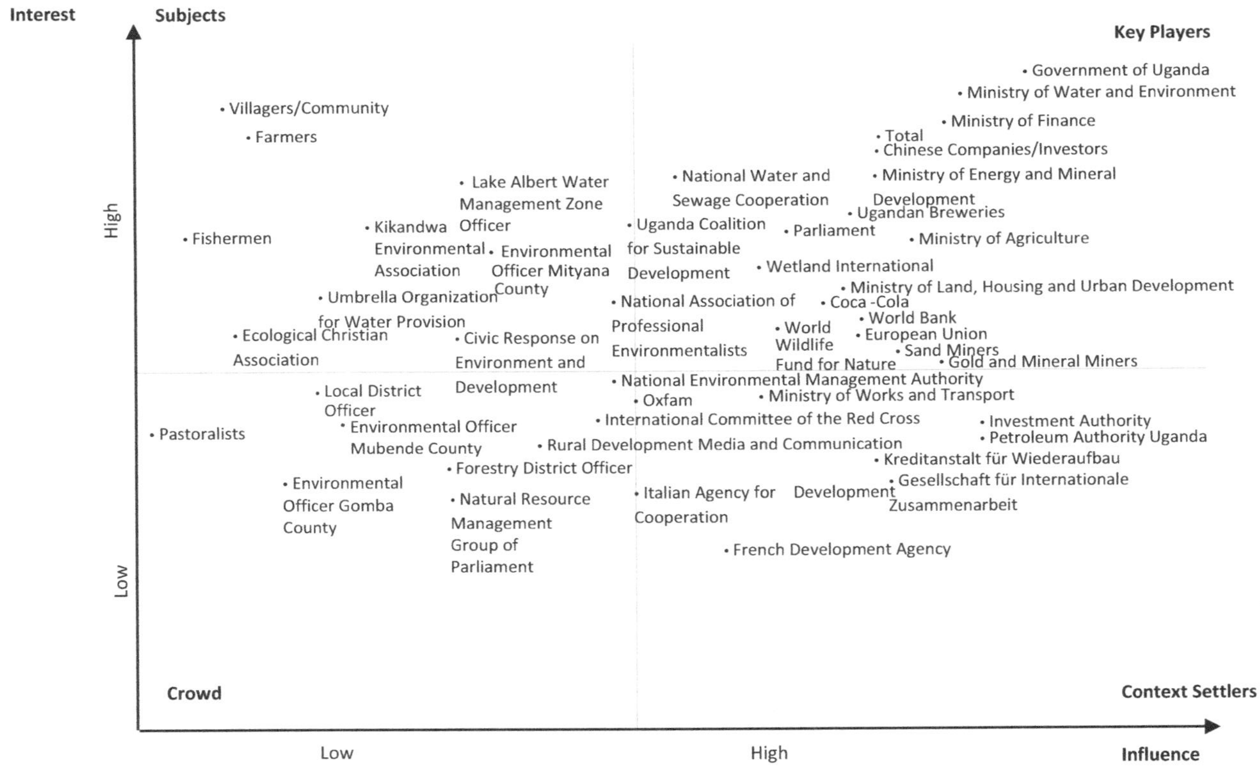

FIGURE 3.4 Influence-Interest Matrix Lake Wamala

(*Source:* The Author, based on Grimble and Wellard 1997)

3.5 Stakeholder Matrix – Interests and Influences

As can be seen from Figures 3.3 and 3.4, the majority of stakeholders are in between subjects and crowd. The majority of stakeholders, however, are leaning towards crowd, with low interest and little or no influence over the water governance and resource management processes.

From the figures, it can be seen that there are six national actors classified as Key Players. The national political actors, in particular the Governments of Kenya and Uganda, the Ministry of Water and Sanitation, the Ministry of Water and Environment and the National Treasury, as well as the Ministry of Finance, have the highest influence on the basin's natural resource management. This is understandable since the governments and the Ministries of Water have the regulatory power and authority which they exercise in managing the basin and in allocating land and water resources. The ministries and departments are charged with the supervision and co-ordination of all matters related to environmental and natural resource governance. This mandate gives them greater influence in the management of the basin and its respective resources. It is apparent from Figures 3.3 and 3.4 that two international economic actors are also highly influential in the area. This stems from their contribution to the economic development not only within Lake Naivasha and Lake Wamala basin, but also in Kenya and Uganda in general. Furthermore, they are in control of the resources. In the case of Lake Naivasha, they improve the livelihood of the local resource users because they offer labour to the villagers. In particular, the flower farms offer labour to the local population and attract people from western Kenya to work in the farms (Kapila 2018). Interestingly, energy and geothermal companies can be classified as Context Settlers, with low interest and high influence. This result seems counterintuitive, as they rely on land and water to produce geothermal energy. The government's focus on economic upgrading processes, however, grants economic companies a preferential treatment and supply with the needed resources. Accordingly, this results in the economic actors' low interest in resource management. Likewise, they are highly influential because the national government depends on their generosity to be able to meet their national economic goals (see Chapter 4). In Uganda, both the Petroleum and Investment Authorities are classified as Context Settlers. The Government of Uganda, as well as the ministries, authorizes the Investment and Petroleum Authorities to allocate the resources to international investors. Thus, their low interest is owing to the national governments specifications of land and water uses. Most of the time, the instructions come directly from the president's office. However, both authorities are highly influential as they are responsible for securing higher dividends through the allocation of investments within the country (Anthony 2019).

Sub-national agencies and departments are foremost located at the edge between Subject and Crowd with low influence and low interest. Their low interest derives from their minimal operational capacities, as both lake basins still continue to experience environmental and resource quality degradations. Even though the Kenyan Water Act of 2016 provides county agencies and departments with more responsibilities (Water Act 2016), sub-national governments and agencies still lack the necessary structures and strategies for sustainable resource management (Bokokamau 2018). Uganda is still in the process of decentralizing its natural resource governance structures. Consequently, neither the Lake Albert Water Management officer nor the three district officers of Mityana, Mubende and Gomba have influences over decision-making processes. Additionally, the set-up of three districts and another water management institution has become another resource in a broader struggle over authority between centre and district because all are responsible for natural resource management. Resulting thereof, the influences of each sub-national department is decreasing.

Across the interviews, the interviewees reported that local non-governmental organizations have low influence over governance processes. The inspection of the aforementioned figures shows that community-based and local non-governmental organizations are classified as in between Subjects and Crowd with low influence and low interest. As pointed out, their interest springs from the mandate to supervise and coordinate matters related to the environment and to oversee the implementation of policies relating to the ecosystem (Anguparu 2019; Mooya 2019). The attempts of these local political actors and NGOs to acquire control over political actions concerning environmental management processes have, however, failed. Thus, this results in their low interest in the water and land resources currently. Local stakeholders expressed concerns that the decline of the community-based organizations' engagement results in a low interest to initiate their own natural resource processes and programs. When asked whether some of the initiatives have been successful, the majority of the respondents highlighted that the basins still experience environmental and water quality degradations despite their efforts to counteract the pollution of the lake's water (Kaganga 2018; Kiwazi 2019; Muigai 2019). Added to that, their attempts to fulfil an adequate natural resource management process also failed, even though the number of NGOs dealing with natural resources is very high. The analysis identified that community-based organizations weaken their level of influence as they do not coordinate themselves or accelerate a closer integration of their working agendas. Hence, concerns can be expressed about their future influence regarding a more sustainable natural resource governance within Lake Naivasha or Lake Wamala basin.

There are two outstanding community-based organizations for Lake Naivasha and Lake Wamala which are Imarisha and Lake Naivasha Flower Council for Lake Naivasha and Wetland International and Uganda Coalition for Sustainable Development for Lake Wamala. From the figures, it can be seen that, combined, they have the highest level of interest and influence. Thus, they are outliers due to their high influences compared to the other community-based and non-governmental organizations. Both have been actively involved in the basins for over 15 years. Imarisha is focusing on catchment protection and the improvement of the resource users' livelihood (Bokokamau 2018). Imarisha has further been successful to get parliamentarian groups and the county government involved in its work and to set environmental issues on the political agenda (Bokokamau 2018). Wetland International is working closely with the Natural Resource Management Group of Parliament concerning wetland protection and strengthening the basin's climate change resilience, albeit, so far without success on a parliamentarian hearing or even a legislative resolution (Kobusingye 2019). Lake Naivasha Flower Council's high influence results from the region's significant role in the international trade of cut flowers and ornamentals. The Flower Council supports the national government in facilitating trade through the provision of incentives to mostly European flower companies (Kenyan Flower Council 2020). Some interviewees argued that flower farms receive zero or reduced duties on water recycling and zero or reduced taxes on imported inputs crucial to the sector (Kapila 2018).

Overall, the participants demonstrated that local resource users' influence regarding natural resources governance processes is close to zero (AGEH 2018; Kapila 2018; Peter 2018; Yakub 2019). The pastoralists, especially the Maasai, perceive themselves as having the least influence while being the most affected (Vivekananda 2015). A former pastoralist reported that small-holder farmers, especially, limit their influence, accusing them to only know about herding, forcefully destroying the small-holder farmers land and that they are without other skills necessary for the regular job market (Balama 2019). '[Pastoralists] are not welcomed' he commented, even though pastoralists move their cattle down to the lakeshores via one of the four remaining public access sites 'only [to] get water [for] one or two days' (Ibid. 2019). Small-scale farmers indicated that their farmland is reduced due to the set-up of economic sites. Even though most residents around the lake still depend on agricultural activities to pursue a living, agriculturalists and others reported that they are excluded from decision-making processes despite their efforts to demonstrate against the closing of farmland and access routes to the lake (Peter 2018; Kimeta 2019). Villagers, especially flower pickers in the flower farms, report that 'they will lose their jobs' (Peter 2018) in the farms if they stand up or raise their voices against the inhuman working conditions, low wages and sexual harassments. Compared to the other local resource users, fishermen have the highest influence over decision-making processes. Most fishermen unified in fishing coalitions to increase their sales market by also selling fish to supermarkets in Naivasha or in Nairobi and other businessmen. As there is currently still enough water in the lake, they use their influence as a fishing association to keep the four landing sites open. These landing sites further serve as marketplaces. However, it was reported that the fishermen do not have influence to limit the discharge of chemicals into the lake directly. These chemicals negatively affect the fish population and result in smaller fish and reduced fish population (Fishing Community Karagita Landing Site 2019). In the case of Lake Wamala, informants agreed that the local population, which is living at the lake's shorelines directly, is decreasing. International and national companies purchase land titles from governmental officials to set up their industrial sites close to the resources. Consequently, these companies take over land which was used by the fishermen, small-holder farmers and villagers previously. Some interviewees demonstrated that local resource users have been compensated for the loss of their land financially (Muhindo 2019), while others argued that relocations are commonly used to compensate the local population (Anthony 2019; Kobusingye 2019). All the same, the number of local resource users decreased. Furthermore, local resource users are living in outlying regions without access to water bodies.

Whereas small-holder farms are still able to do farming in these new areas, fishermen are no longer able to fish. Relocations took place without taking into consideration 'alternative sources of water for those communities' to pursue a living (Twesigye 2019).

In summary, the SA demonstrated that there is a big discrepancy between the influence actors should have given the institutional framework and the actual influence on natural resource management. Together, these results also provide important insights into the geographical arrangement of the actors around the lake. Based on the observations, the interviews and the analysis, a mapping of both lake basins was created portraying the location of the various stakeholders (see Figure 3.5).

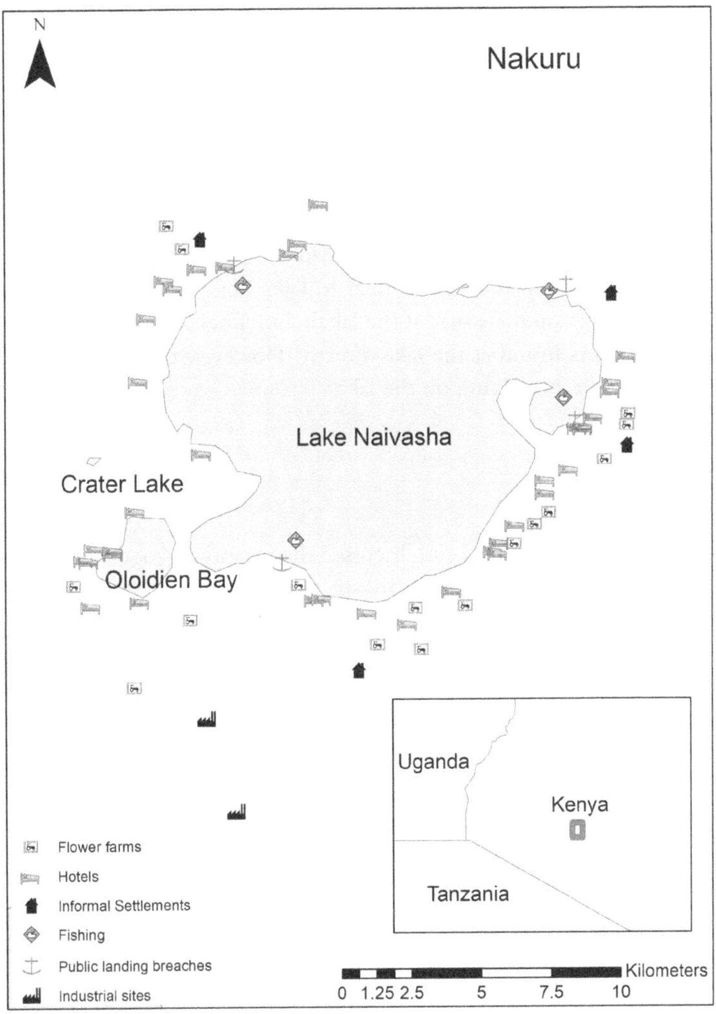

FIGURE 3.5 Geographical Location of the Stakeholders Around Lake Naivasha

(*Source:* Shah for the Author 2019)

Mapping Lake Naivasha, it can be seen that the international and national economic actors (flower farms and hotels) are mainly located directly at the lake's shorelines.

Furthermore, the industrial and geothermal sites are placed slightly above the lake as they rely on the volcanic rocks to pursue their businesses. It is apparent from the map that the northern and north-western shorelines are sparsely inhabited by flower farms. As indicated at the beginning of this chapter, descendants of former white settlers reside there. Furthermore, the ground is more fertile in the southern and south-eastern parts of the lake, thus better suitable for agriculture and horticulture. A closer inspection of the map shows that flower farms and informal settlements cannot be found around Lake Oloidien. As pointed out at the beginning of this chapter, Lake Oloidien is not a freshwater lake. It is, consequently, unsuitable for agricultural and horticultural activities. Due to the decreasing availability of land around Lake Naivasha, hotel companies set up their premises around Lake Oloidien. The lake is advertised as a flamingo viewing hotspot by hotels and tourist companies (Peter 2019). Lastly, local resource users live in one of the four informal settlements which are located outside the landing sites. From the map of Lake Wamala (see Figure 3.6), it can be seen that the group of local resource users is not found at the lake anymore. While there were four local fishing villages in 2018, only three remained open in 2019. The map shows that local resource users are located father away from the lake. Similar to Lake Naivasha, national and international companies set up their sites at the lake's shorelines directly. In contrast, there are less stakeholders found at the lake directly. However, the number of actors in the surrounding region relying on the lake's resources exceeds the number around Lake Naivasha.

3.6 Brief Summary

Firstly, this chapter has charted and discussed the location, geographical and hydrological characteristics of both Lake Naivasha and Lake Wamala. Secondly, the chapter analysed the various stakeholders living and working around the two lakes in reference to their interests in the water and land resources. The analysis indicated that both lakes have emerged as economic hotspots. The main segments of the broader economic processes comprise sand mining, rice growing and oil explorations around Lake Wamala. At Lake Naivasha in particular, horticulture, geothermal energy and tourism are dominating the basin's economic activities. At the same time, the amount of people relying either on the lake's resources directly or indirectly has increased over the course of the past 20 years. Whereas the amount of people residing in close proximity to the lake increased in the case of Lake Naivasha, the number of residents at Lake Wamala decreased. Moreover, the total population living further away from Lake Wamala's shorelines increased. As a result of insufficient water resources, they still depend on the lake's resources to pursue a living.

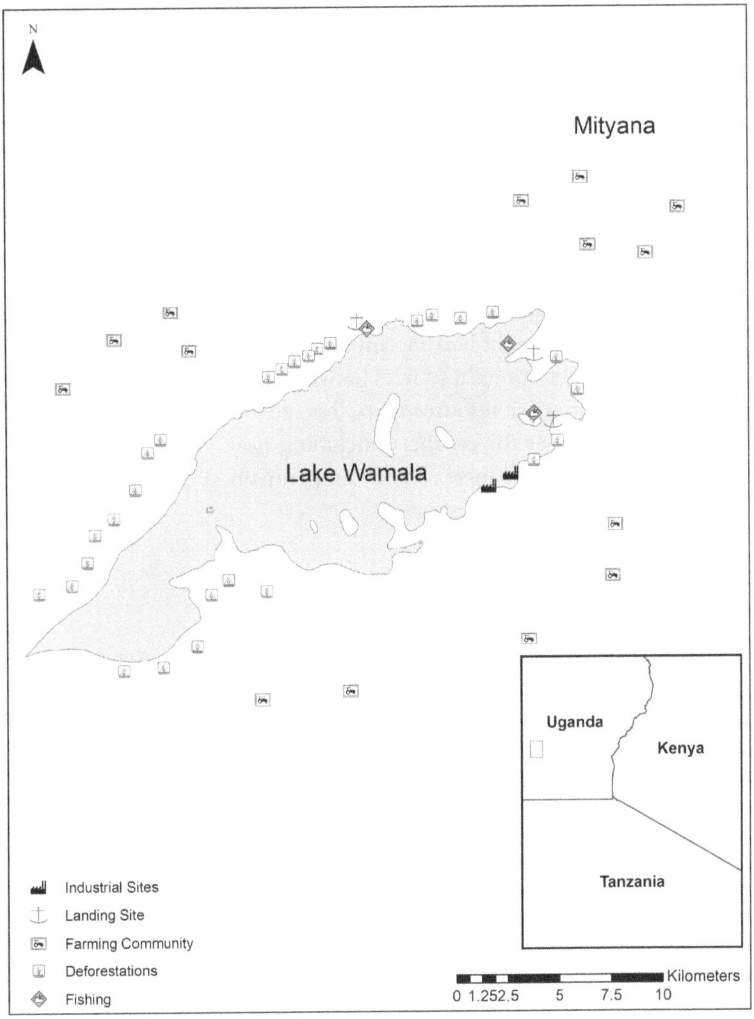

FIGURE 3.6 Geographical Location of the Stakeholders Around Lake Wamala

(Source: Shah for the Author 2019)

The second part of the chapter discussed the influences on integrating natural resource governance in the institutional framework of the identified actors which are involved in the use of the lake's water and land resources. The section answered the question of which influences each group of actors have on the institutional framework to allow for a better distribution and accessibility of the lake's resources. It critically assessed the linkages between the interests of the various stakeholders and their influences on water management processes. The institutional framework contextualizing water management is not reflected in the influences of the individual stakeholders with reference to their interests. Consequently, decision-making processes with regard to water and land are significantly shaped by national political actors. In addition to regulative measures, national and international economic actors are increasingly influential in using public resources to increase their economic investments. Overall, these results indicate that the different standards and contributions to the political elites' strategic priorities reflect the powerful role of gaining influence over resource access and water management respectively. Those who hold powerful and profitable positions are further reflected in their location and proximity to the lakes and its resources.

The results in this chapter indicate that the competing interests and influences over water management might become another resource in a broader struggle over access to and the distribution of resources between the different stakeholder groups but also within one group of resource users. The next chapter, therefore, moves on to discuss possible reasons for conflicts, including how political, socio-economic and environmental processes increase the competitiveness over the available water resources and the social outcomes thereof.

Bibliography

AGEH Employee (2018, 1 August): Interview by Renner, J. Nairobi, Kenya.

Anders, D. (2018, 20 August): Interview by Renner, J. Kampala, Uganda.

Anguparu, M. (2019, 24 September): Interview by Renner, J. Kampala, Uganda.

Anthony (2019, 6 August): Interview by Renner, J. Kampala, Uganda.

Bakari (2018, 15 August): Interview by Renner, J. Nairobi, Kenya.

Baker, M., Sarfo, I., Darko, G. and Bi, S. (2019): Loss of wetland resources in Uganda: The case of Wamala in Mityana District. *International Research Journal of Public and Environmental Health* 6 (8), 170–190.

Balama, J. (2019, 15 August): Interview by Renner, J. Nairobi, Kenya.

Bbira, Y. (2018, 24 August): Interview by Renner, J. Mityana, Uganda.

Becht, R., Odada, E. O. and Higgins, S. (2006): *Lake Naivasha: Experiences and Lessons Learnt Brief.* International Lake Environment Committee, Lake Basin Management Initiative. Enschede.

Bokokamau (2018, 9 August): Interview by Renner, J. Naivasha, Kenya.

Carolina, B. F. (2002): *Competition over Water Resources: Analysis and Mapping of Water-related Conflicts in the Catchment of Lake Naivasha (Kenya).* ITC. Eschede.

Clark, C. E., Harto, C. B., Sullivan, J. L. and Wang, M. Q. (2011): *Water Use in the Development and Operation of Geothermal Power Plants.* Argonne National Laboratory ANL/EVS/R-10/5. Oak Ridge.

Community Members Karagita Landing Site (2018, 11 August): Interview by Renner, J. Karagita Landing Site, Kenya.

European Commission (2020): *Global Surface Water Explorer.* At: https://global-surface-water.appspot.com/ [retrieved last 8 May 2020].

European Commissions's Joint Research Centre (2019): *Global Surface Water Explorer. Lake Naivasha.* At: https://global-surface-water.appspot.com/map [retrieved last 19 May 2019].

Fauvel, J.-B. (2019, 19 July): Interview by Renner, J. Kampala, Uganda.

Fishing Community Karagita Landing Site (2019, 20 August): Interview by Renner, J. Karagita Landing Site, Kenya.
Flower Farm (2019, 21 August): Interview and Company Visit by Renner, J. Lake Naivasha, Kenya.
Gesellschaft für Internationale Zusammenarbeit (2018): *Uganda: Enhanced Water Security and Sanitation Programme (ENWASS)*. Deutsche Gesellschaft für Internationale Zusammenarbeit. Eschborn.
Grimble, R. and Wellard, K. (1997): Stakeholder methodologies in natural resource management: A review of principles, contexts, experiences and opportunities. *Agricultural Systems* 55 (2), 173–193.
Harper, D. M., Boar, R. R., Everard, M. and Hickley, P. (2002): *Developments in Hydrobiology: Lake Naivasha, Kenya*. Kluwer Academic Publishers. Dordrecht.
Harper, D. M., Mavuti, K. M. and Muchiri, S. M. (1990): Ecology and management of Lake Naivasha, Kenya, in relation to climatic change, alien species introductions and agricultural development. *Environmental Conservation* 17, 328–335.
Harper, D. M., Morrison, E. H. J. and Macharia, M. M. (2011): Lake Naivasha, Kenya: Ecology, society and future. *Freshwater Reviews* 4 (2), 89–114.
International Panel on Climate Change (2007): *Climate Change 2007. Climate Change Impacts, Adaptation and Vulnerability*. Cambridge University Press. Cambridge.
Juenger, O. (2018, 20 August): Interview by Renner, J. Kampala, Uganda.
Kaganga, J. (2018, 27 August): Interview by Renner, J. Kampala, Uganda.
Kaganga, J. (2019, 23 July): Interview by Renner, J. Kikandwa, Uganda.
Kamese, G. (2018, 17 August): Interview by Renner, J. Kampala, Uganda.
Kapila, S. (2018, 10 August): Interview by Renner, J. Lake Naivasha, Kenya.
Kenyan Flower Council (2020): *Kenyan Flower Council. Quality Assured*. At: http://kenyaflowercouncil.org/?page_id=96 [retrieved last 18 January 2020].
Kikandwa Community Member (2019, 23 July): Interview by Renner, J. Kikandwa, Uganda.
Kimbowa, R. (2018, 20 August): Interview by Renner, J. Kampala, Uganda.
Kimeta, E. (2018, 10 August): Interview by Renner, J. Naivasha, Kenya.
Kimeta, E. (2019, 19 August): Interview by Renner, J. Naivasha, Kenya.
Kitamirike, J. (2019, 23 August): Interview by Renner, J. Kampala, Uganda.
Kitzio, G. (2019, 22 July): Interview by Renner, J. Kikandwa, Uganda.
Kiwazi, E. (2019, 24 July): Interview by Renner, J. Kikandwa, Uganda.
Kobusingye, L. (2019, 5 August): Interview by Renner, J. Kampala, Uganda.
Kuiper, G. and Gemählich, A. (2017): Sustainability and depoliticisation: Certifications in the cut-flower industry at Lake Naivasha, Kenya. *Africa Spectrum* 52 (3). 31–53.
Kundu, R., Mulanda, A. C., Muchiri, M., Njiru, J. H. and Ojuok, J. E. (2010): Difficulties of fishing at Lake Naivasha, Kenya: Is community participation in management the solution? *Lake & Reservoirs: Research and Management* 15 (1), 15–23.
Liebig, K. (2020, 23 January): Interview by Renner, J. Frankfurt am Main, Germany.
Mbaisi, C. (2018, 15 August): Interview by Renner, J. Nairobi, Kenya.
Mekonnen, M. M., Hoekstra, A. Y. and Becht, R. (2012): Mitigating the water footprint of export cut flowers from the Lake Naivasha Basin, Kenya. *Water Resource Management* 26, 3725–3742.
Mooya, B. (2019, 25 September): Interview by Renner, J. Kampala, Uganda.
Moses (2019, 8 August): Interview by Renner, J. Kampala, Uganda.
Mucunguzi, S. (2019, 30 July): Interview by Renner, J. Kampala, Uganda.
Muhindo, J. (2019, 31 July): Interview by Renner, J. Kampala, Uganda.

Muigai, P. (2019, 21 August): Interview by Renner, J. Naivasha, Kenya.

Mulatu, D. W., van Oel, P. R. and van der Veen, A. (2015): Firms' willingness to invest in a water fund to improve water-related ecosystem services in the Lake Naivasha basin, Kenya. *Water International* 40 (3), 463–482.

Musinguzi, L., Efitre, J., Odongkara, K., Ogutu-Ohwayo, R., Muyodi, F., Natugonza, V., Olokotum, M., Namboowa, S. and Naigaga, S. (2015): Fishers' perceptions of climate change, impacts on their livelihoods and adaptation strategies in environmental change hotspots: A case of Lake Wamala, Uganda. *Environment, Development and Sustainability* 18, 1255–1273.

Mutambukah, B. (2019, 29 July): Interview by Renner, J. Kampala, Uganda.

Nakaggwa, R. (2019, 29 July): Interview by Renner, J. Kampala, Uganda.

Namukose, M. (2019, 26 September): Interview by Renner, J. Kampala, Uganda.

National Geographic (2020): *Rift Valley*. At: www.nationalgeographic.org/encyclopedia/rift-valley/ [retrieved last 12 January 2020].

Natugonza, V. and Ogutu-Ohwayo, R. E. J. (2015): The responses of Nile Talapia (Linnaeus, 1785) in Lake Wamala (Uganda) to changing climatic conditions. *Lakes and Reservoirs: Resources and Management* 20 (10), 101–119.

Ogada, J. O., Krhoda, G., van der Veen, A., Marani, M. and van Oel, P. (2017): Managing resources through stakeholder networks: Collaborative water governance for Lake Naivasha basin, Kenya. *Water International* 42 (3), 271–290.

Olupot, W. (2018, 22 August): Interview by Renner, J. Kampala, Uganda.

Peter, S. (2018, 9 August): Interview by Renner, J. Naivasha, Kenya.

Peter, S. (2019, 19 August): Interview by Renner, J. Naivasha, Kenya.

Ramsar (2011): *National Report on the Implementation of the Ramsar Convention on Wetland*. National Reports to be submitted to the 11th Meeting of the Conference of the Contracting Parties, Romania.

Ran, A. M. (2018, 14 August): Interview by Renner, J. Nairobi, Kenya.

Republic of Kenya (2019): *2019 Kenya Population and Housing Census. Volume I: Population by County and Sub-County*. Republic of Kenya. Nairobi.

Schäfer, D. (2018, 14 August): Interview by Renner, J. Nairobi, Kenya.

Standardised Precipitation-Evapotranspiration Index (2021a): *Drought Monitor Lake Naivasha*. At: https://spei.csic.es/map/maps.html#months=1#month=11#year=2019 [retrieved last 15 November 2021].

Standardised Precipitation-Evapotranspiration Index (2021b): *Drought Monitor Lake Wamala*. At: https://spei.csic.es/map/maps.html#months=1#month=11#year=2019 [retrieved last 15 November 2021].

Twesigye, B. (2019, 6 August): Interview by Renner, J. Kampala, Uganda.

United Nations Education, Scientific and Cultural Organization (1999): *Lake Naivasha*. At: https://whc.unesco.org/en/tentativelists/1345/ [retrieved last 16 November 2019].

United Nations Environmental Programme (2009): *Atlas of Our Changing Environment*. UNEP. Nairobi.

USAID (2016): *Development of Kenya's Power Sector 2015–2020*. USAID. Nairobi.

Van Oel, P., Mulatu, D. W., Odongo, V., Meins, F. M., Hogeboom, R. J., Becht, R., Stein, A., Onyando, J. and Van der Veen, A. (2013): The effects of groundwater and surface water use on total water availability and implications for water management: The case of Lake Naivasha, Kenya. *Water Resources Management* 27 (9), 3477–3492.

Vergin, J. (2019, 8 October): Interview by Renner, J. Kampala, Uganda.

Vianny, N., Olokotum, M. and Musinguzo, L. (2016): Implications of climate warming for hydrology and water balance of small shallow lakes: A case of Lake Wamala and Kawi, Uganda. *Aquatic Ecosystem Health and Management* 19 (4), 327–335.

Vivekananda, J. (2015): *Peace Audit Kenya.* International Alert. London.

Water Act (2016): *The Water Act 2016.* National Council for Law Reporting. Nairobi.

Willy, D. K., Zhunusova, E. and Holm-Müller, K. (2014): Estimating the joint effect of multiple soil conservation practices: A case study of smallholder farmers in the Lake Naivasha basin, Kenya. *Land Use Policy* 39, 177–187.

World Bank Group (2018): *Kenya Launches Ambitious Plan to Provide Electricity to All Citizens by 2022.* At: www.worldbank.org/en/news/press-release/2018/12/06/kenya-launches-ambitious-plan-to-provide-electricity-to-all-citizens-by-2022 [retrieved last 10 January 2020].

World Bank Group (2019): *Creating Markets in Kenya. Unleashing Private Sector Dynamism to Achieve Full Potential.* World Bank Group. Washington.

World Wildlife Fund Employee (2018, 10 August): Interview by Renner, J. Naivasha, Kenya.

Yakub, Y. (2019, 2 September): Interview by Renner, J. Nairobi, Kenya.

4

THE NEXUS OF POLITICAL AND ECONOMIC VULNERABILITY, ADAPTATION AND RESILIENCE

4.1 Vulnerability and Adaptive Capacity in Lake Naivasha and Lake Wamala basin

To understand why the risk over the outbreak of a conflict is higher among the local-level actors compared to all other identified ones, this shall be analysed using the nexus of vulnerability and adaptive capacity to cope with external changes. The approach is based on the assumption that human welfare increases environmental changes associated with climate change and increased natural hazards. Therefore, vulnerability encompasses the two elements, nature and society, as a coupled system (Raleigh et al. 2008). Hence, the aspects that affect vulnerability need to be explained using a combination of variables, including social factors and environmental risks (i.e. physical aspects of climate related events). Risk for the purpose of this book is understood as the product of the impact of a climatic event and vulnerability (Scheffran et al. 2012).

Vulnerability is defined as the ability or inability of individuals or social groups to adapt to 'any external stress placed in their livelihood and well-being' (Kelly and Adger 2000: 327). In relation to forms of co-ordination, the existing literature disaggregates vulnerability into the two distinct aspects of individual and collective vulnerability. This distinction allows to clarify the scale issue and the unit of analysis better (Adger 1999). Accordingly, individual vulnerability is determined by the social status of individuals, resource access or the diversity of incomes. On the other hand, collective vulnerability deals with nations, regions or communities. It identifies infrastructure, or institutional and market structures (Ibid. 1999).

The existing literature on vulnerability explains social vulnerability as an environmental condition which is fundamentally for adaptive capacity of the economic and institutional context (Methmann and Oels 2014). As a result, the socio-economic context and, in particular, the impact of the economic agenda on individuals and

DOI: 10.4324/9781003271291-4

stakeholder groups' poverty levels, resource dependency and inequality are considered. Poverty can be seen as an important aspect of social vulnerability because it is indicated through the quantifiable indicator of income. Income can serve as another proxy for access to resources. According to Blaikie et al. resource access can be defined as 'the ability of an individual . . ., group or community to use resources which are directly required to secure a livelihood' (Blaikie et al. 1994: 48). The dependency of individuals on resources is constituted by the reliance on a narrow range of resources leading to social and economic stability. Increased stresses and increased external influences on these resources result in higher risks of social and economic instability. Additionally, vulnerability to climate extremes is determined by formal institutional arrangements. Consequently, if institutions fail to plan for changing governmental or climatic conditions, the vulnerability of individuals or groups increases (Adger 2006).

Therewith, access to and the distribution of resources is based on the social and economic agenda and the infrastructural and institutional arrangements to limit the exposure to multi-facetted risks. The following two sub-sections take a closer look at climate data and how the research areas are affected by climate changes. They further focus on how the economic transformation agendas and the institutional effectiveness of both countries influence resource access and resource distribution.

4.2 Temperature and Precipitation Deviations Across Kenya and Uganda

Kenya's and Uganda's economy still rely heavily on the agricultural sector. On average, agriculture contributes 30 per cent to both countries' GDP. As a mean, the sector employs 65 per cent of the total population (CIA 2022a, 2022b) and more than 75 per cent of Kenya's and 70 per cent of Uganda's rural people (FAO 2018, 2020; World Bank Group 2020a). The pastoral communities and smallholder farmers especially depend almost exclusively on rain-fed pasture and water for their survival and for their domestic use (Njiru 2012; see Table 4.1).

Therewith, the agricultural sector is extremely vulnerable to climatic changes and the impacts of natural hazards (Ibid. 2002). This, in return, increases the contextual vulnerability of local resource users to resource access and, therewith,

TABLE 4.1 Sensitivity of Population Based on the Importance of Agricultural Activities

	Water Withdrawals	*Rain-Fed*	*Agriculture*	
	% used for agriculture of total water withdrawal	*% of total agricultural area*	*% of gross domestic product*	*% of labour force*
Kenya	80.21 (2016)	75 (2016)	34.5 (2017)	61.1 (2015)
Uganda	40.66 (2008)	80 (2017)	28.2 (2017)	71.0 (2013)

(*Source*: The Author 2020, based on FAOstat 2017, Aquastat 2020 and CIA 2022a, 2022b)

resource distribution. As indicated earlier, the temperature in both Kenya and Uganda has risen by roughly 1°C over the past 20 years and is expected to increase further. It is projected to rise by 1.5°C until 2040 (Mukwaya et al. 2012).

Furthermore, both countries experienced an increase in rainfall activities as a result of global climate change. Especially in the arid areas, the rise in temperatures resulted in the drying up of riverbeds or reduced water levels. A much-debated question in the literature is whether climate change is a major driver of water shortages and conflicts or if other aspects of the socio-economic and political environment drive conflict dynamics (see, e.g. Thomas 2008; Schmitz et al. 2013; Scheffran et al. 2019). Thus, studies provide hydrological scenarios on the scale of different ecological zones in Kenya and Uganda (see, e.g. International Research Institute for Climate and Society 2020; SPEI 2020).

An objective of the study was to investigate the temperature and precipitation differences between the research sites and the arid areas of Turkana and Karamoja to determine whether climate change affects resource shortages. Figures 4.1 and 4.2 are based on climate data received from the University of East Anglia and the World Bank Group. Both figures present the deviations of the annual temperature and precipitation from the arithmetic means over the past 20 years for Lake Naivasha and Lake Wamala. The continuous lines show the mean values of temperature and

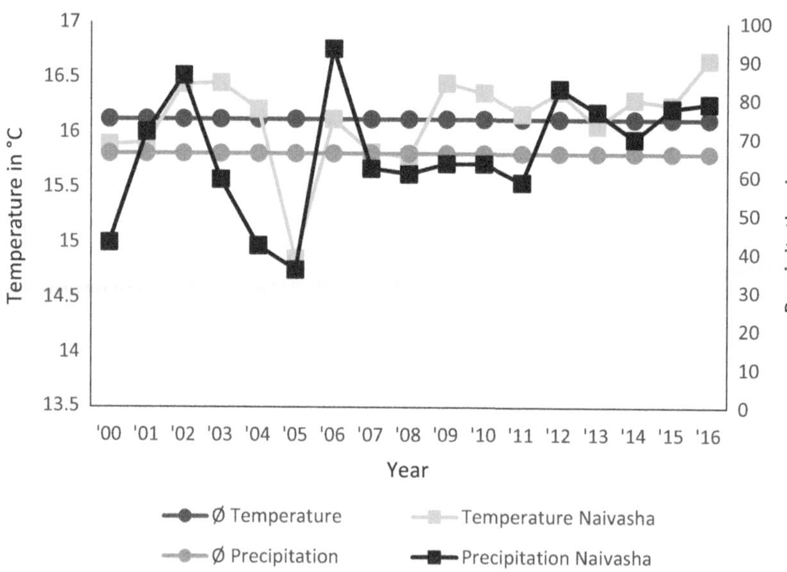

FIGURE 4.1 Temperature and Precipitation Deviation Lake Naivasha

(Source: The Author 2020)

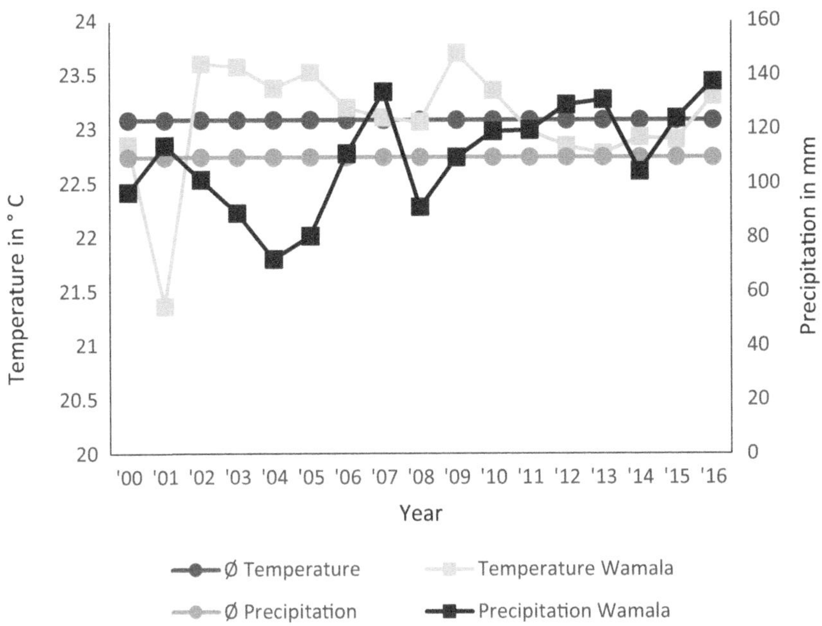

FIGURE 4.2 Temperature and Precipitation Deviation Lake Wamala

(*Source:* The Author 2020)

precipitation over the past 20 years. The other two curves illustrate the annual temperature and precipitation.

Taking a closer look at the data for Lake Naivasha, the mean annual temperature has been above the last 20 years' average since 2009. The temperature for Lake Naivasha has risen by a mean value of almost 0.4°C over the past ten years. Naivasha has experienced a major drought in 2005. After that, the annual rainfall has recovered to the 20 years' average. Since 2011, the area has experienced more than the average rainfall. However, the main characteristic of the increase in precipitation is an increase in heavy rainfall, which the area experiences over a short period of time, as well as longer and more intense heatwaves. Compared to Lake Naivasha, the mean temperature and mean precipitation are higher for Lake Wamala.[1] Similarly, the temperatures are also increasing for the Lake Wamala area, while, at the same time, rainfall has become more intense over the past 18 years.

The results from the field visits to the arid areas indicate an increase of Turkana and Karamoja's mean temperature by 4°C, while the simulations converge towards a decrease in precipitation (Government of Kenya 2018; World Bank Group 2020b). As a consequence of the decrease in rainfall, the water levels of water bodies decrease. This results in a significant decline in water supply to dams and reservoirs. Furthermore, extreme weather events intensify and put additional

TABLE 4.2 Synthesis of Temperature and Precipitation Patterns

	Mean Temperature	*Extreme Temperature / Heat Waves*	*Mean Precipitation*	*Heavy Rainfall*	*Drought*
Lake Naivasha	+	+	-	+	+
Turkana	++	+++	--	++	+++
Lake Wamala	+	++	-	+	++
Karamoja	++	+++	--	++	+++

(*Source*: The Author 2020 based on Opiyo et al. 2014; Government of Kenya 2018; World Bank Group 2020c; International Research Institute for Climate and Society 2020; SPEI 2020)

Semi-quantitative evaluation of mean temperature and mean precipitation in absolute terms of ecological zone 5 (Turkana and Karamoja) and ecological zone 3 (Lake Naivasha and Lake Wamala). + > 2°C; ++ < 2°C to > 4°C; +++ < 4°C compared to the 20th century. Precipitation and extreme weather events (heat waves, heavy rainfall and drought) are shown in relative terms: + indicates an increase by 10 per cent compared to the 20th century; ++ indicates an increase by 20 per cent and; +++ by < 20 per cent. – indicates a reduction in rainfall by 10 per cent compared to the 20th century; – indicates a reduction in rainfall by 20 per cent; – by < 20 per cent. However, a reduction does not necessarily entail.

stress on the water availability in these areas. It was observed that River Kawalasee in Lodwar dries up completely during the dry season as a result of the increasing temperatures. This finding is supported by the research findings of the Turkana Observatory Research Centre that analyses the impacts of climate change on the water availability in Turkana (Ongech 2019).

However, neither Lake Naivasha nor Lake Wamala is located in the mentioned arid areas (see Chapter 3). The results from the climate analysis indicate an increase in temperature and heat waves, linked to a rise in shorter but more extreme precipitation for the mentioned areas. Interviewees reported that climate change is affecting the water levels of Lake Naivasha and Lake Wamala, but more moderately than in the case of Turkana and Karamoja (Government of Kenya 2018; World Bank Group 2020c; see Table 4.2).

Respondents underlined that the decline in surface runoff at Lake Naivasha and Lake Wamala is not due to climate change (Opiyo 2018; Kiboye 2018; Kamese 2018). Instead, they argued that economic infrastructural projects increase the pressure on the lakes' water resources. Thus, economic and environmental projects accelerate the climate effects on the lake's water and land resources (Gumba 2018; Kamese 2018; Konbusingye 2019). These include deforestations, soil degradations or wetland destructions. As mentioned, both Lake Naivasha and Lake Wamala experienced an increase in economic infrastructural projects.

4.3 Economic Development Agendas of Kenya and Uganda

In its national economic development agenda, Kenya has committed itself to become a middle-income country by 2030. In order to achieve this aim, the

Government of Kenya (GoK) created the Kenya Vision 2030 to create 'a globally competitive and prosperous country with high quality of life by 2030' (Kenya Vision 2020). To realize this vision, Kenya has set up the Big 4 development agenda covering four sectors. The four sectors include the economic and macro pillar, the social pillar, the political pillar and the foundations for the pillars[2] (Ibid.). To achieve the status of a middle-income country, the GoK, furthermore, has set up Medium Term Plans (MTPs) which shall guide the roadmap to achieve the aims of Kenya Vision 2030. The first Five-Year Plan (2008–2012) aimed to implement Flagship Projects, key policies and programmes as defined by the Vision 2030. The flagship projects aimed to achieve higher and sustainable growth of the economy, accompanied by an increase in employment opportunities (Ministry of Finance 2015). During the first five years, the Kenyan government concentrated foremost on the economic and macro pillar. The government intended to raise the ratio of investments to GDP by 10 per cent points,[3] coinciding with a rise of the medium-term domestic product growth per capita to 6 per cent by 2020[4] (Ibid. 2015; World Bank Group 2019).

The second MTP (2013–2017) targeted both the economic and political pillar. With reference to the economic pillar, the vision aimed to steer the GDP growth rate by an average of 10 per cent by 2017 annually. Moreover, the second MTP gave priority to increase the scale and pace of economic transformation through infrastructural development (Ministry of Finance 2015). Under the political pillar, the government gave priority to the implementation of devolution, as spelt out in the constitution (Constitution of Kenya 2010). The third MTP (2018–2022) prioritizes the continuation of the achievements of the first and second MTP. It outlines the concentration on the main policies, as well as legal and institutional reforms. The president, furthermore, focuses on the implementation of, what he calls, the 'Big Four' initiative. The Big Four agenda centres on manufacturing, universal healthcare, affordable housing and food security. It intends to increase the manufacturing share of GDP from 7.7 per cent in 2018[5] to 15 per cent by 2030 (World Bank Group 2020f). In terms of food security, Kenya aims to achieve at least 50 per cent of total agriculture output through agro-processing. Finally, it plans to provide 500,000 units of affordable housings until 2022 and to achieve 100 per cent universal health coverage by 2030. The aims of the economic development agendas and the MTPs are very ambitious. However, interim results suggest that these aims are unrealistic and that the country is unable to achieve the set targets.

To achieve the previously mentioned targets, Kenya, together with international donors, invested more than 7 billion USD in infrastructural projects; more than 176.8 million USD in manufacturing, agro-businesses and services; and around 250 million USD in affordable housing and social projects, like reducing poverty or increasing the level of education, for example (World Bank Group 2019b). As of the end of December 2017, Kenya shifted the focus of its economic agenda. It concentrates on two major projects in the energy sector. The larger of these two projects is the construction of an oil pipeline from Turkana to Mombasa (Ibid. 2019). The construction of a cargo centre within Hell's Gate National Park is the other large infrastructural project (Ran 2018; Kimeta 2019). Hell's Gate National

Park is located at the outskirts of Lake Naivasha. Whereas the oil pipeline is supported by Chinese investors financially, the World Bank Group is the main financial contributor to the cargo centre at Lake Naivasha (World Wildlife Fund 2018; Kimeta 2019; Peter 2019).

Uganda's national economic development strives to reach upper-middle-income status by 2040. The Vision 2040 is conceptualized around strengthening the fundamentals of the national economic system 'to harness the opportunities around the country' (LDPG 2020). The Vision 2040, furthermore, identifies nine key areas, including oil and gas, minerals, ICT[6] business, abundant labour force, trade, water resources, industrialization and agriculture (National Planning Authority 2020). Similarly, the government will implement six National Development Plans (NDP) to achieve the set targets. Two of them have already been formulated. The National Development Plan I (2010/11–2014/15) means to transform the Ugandan society from a 'peasant-based economy' to a peaceful and prosperous middle-income country (Republic of Uganda 2010). Accordingly, the government seeks to improve development indicators closely associated with transformation, including an increase in GDP per capita, raising the human development indicators or improving the level of competitiveness (Ibid. 2010). NDP I is to accelerate socio-economic transformation and to decrease the poverty levels by one third. The plan outlines the improvement of stock and the quality of economic infrastructure. Additionally, NDP I shall strengthen political, economic and corporate governance as well as to promote science, technology and ICT. Among others, NDP I argues that the impacts of macro-economic projects on environmental and climate aspects need to be integrated into the development plans. The plan acknowledges that without climate awareness a successful socio-economic transformation is impossible (Ibid. 2010). However, the plan does not provide any detailed information how any economic transformation can be realised.

The NDP II (2015/16–2019/20) is based on the foundations laid out in the NDP I. The NDP II is to strengthen the country's competitiveness for sustainable wealth creation, employment and inclusive growth. In line with the president's strategic priorities for 2016 to 2021 (State House 2016), the plan stresses the importance to focus on macroeconomic infrastructural projects, human capital development, minerals, oil and gas as well as agriculture and tourism (Republic of Uganda 2015). In his strategic priorities for 2016–2021, the president announced an increase in the investments in health and education, as well as strategic agricultural land, urbanization and transportation (State House 2016). To achieve these targets by the end of the second Five-Year Plan, four core elements have been formulated which include an increase in sustainable production and productivity, an increase in strategic infrastructure to accelerate Uganda's competition, to enhance human capital development, as well as to strengthen mechanisms for quality, effective and efficient service delivery (Republic of Uganda 2015). Contrary to NDP I, NDP II quantifies its targets. After the successful implementation of the NDP II, the growth rate is expected to stand at 6.3 per cent[7] and the GDP per capita income should be at 1,039 USD by the end of 2022.[8] However, due to the long lockdown and the economic stagnation due to the COVID-19 pandemic, this target has not been achieved and it is unlikely that this aim will be achieved any time soon. Furthermore, the president's strategic priorities focus on adding value to raw materials through investments in major oil and gas infrastructures, including the construction of refineries and pipelines. Finally, the Statehouse emphasizes aggressive investments in the tourism sector through marketing, diversification of products and supporting infrastructure and services. The aim is to increase the contribution of tourism to 15 per cent of GDP in 2020, rising by 6 percentage points since 2013 (Republic of Uganda 2015; State House 2016). Contrary to the first plan, climate change and the environmental impact of undertaken infrastructural projects is not acknowledged as a priority anymore (Ibid. 2016). Adding to that, the upcoming NDP III (2020/21–2024/25) will most likely concentrate on growth[9] opportunities in both the oil and gas, and tourism sector. Furthermore, it aims to strengthen the stock and quality of productive infrastructure. This shall coincide with an increase in the inclusiveness and wellbeing of the population, as well as strengthening of the private sector to drive growth and to create jobs (Ministry of Finance, Planning and Economic Development 2020). Overall, both economic development plans are vague and do not provide any clear information on the countries' strategic economic developments. Therefore, it is difficult to quantify if the economic development has been successful or not, so far. Nevertheless, the analysis of the economic development plans showed that macro-economic projects (e.g. infrastructure, transportation and the set-up of industrial sites) are prioritized by both governments regardless of the current economic and social situation.

4.4 Economic Development and Social, Contextual and Governmental Vulnerability

The implementation of Vision 2030 and Vision 2040 is greatly affecting both Lake Naivasha and Lake Wamala. Figure 4.3 summaries identified preliminary effects of the economic processes on the people's vulnerability to economic, social and environmental change at both lakes. The chart shows the effects of the economic agenda on the water and land resources at the two case study locations on the one hand, and the vulnerability of the people living around the lakes to the impacts of the economic agenda on the other hand. It further summarizes how governmental structures catalyse the adaptive capacity of the local resource users. Each element of the economic agenda, the effects of the agenda on the social and contextual vulnerability and adaptive capacity is worded to have the same sign as its category. Major current and future effects are shown in bold. Likely future effects to increase vulnerability are shown dashed and factors reducing vulnerability are presented dotted.

4.4.1 Sensitivity

Around Lake Naivasha, the national government allocated land to national and international businesses to increase infrastructural development through

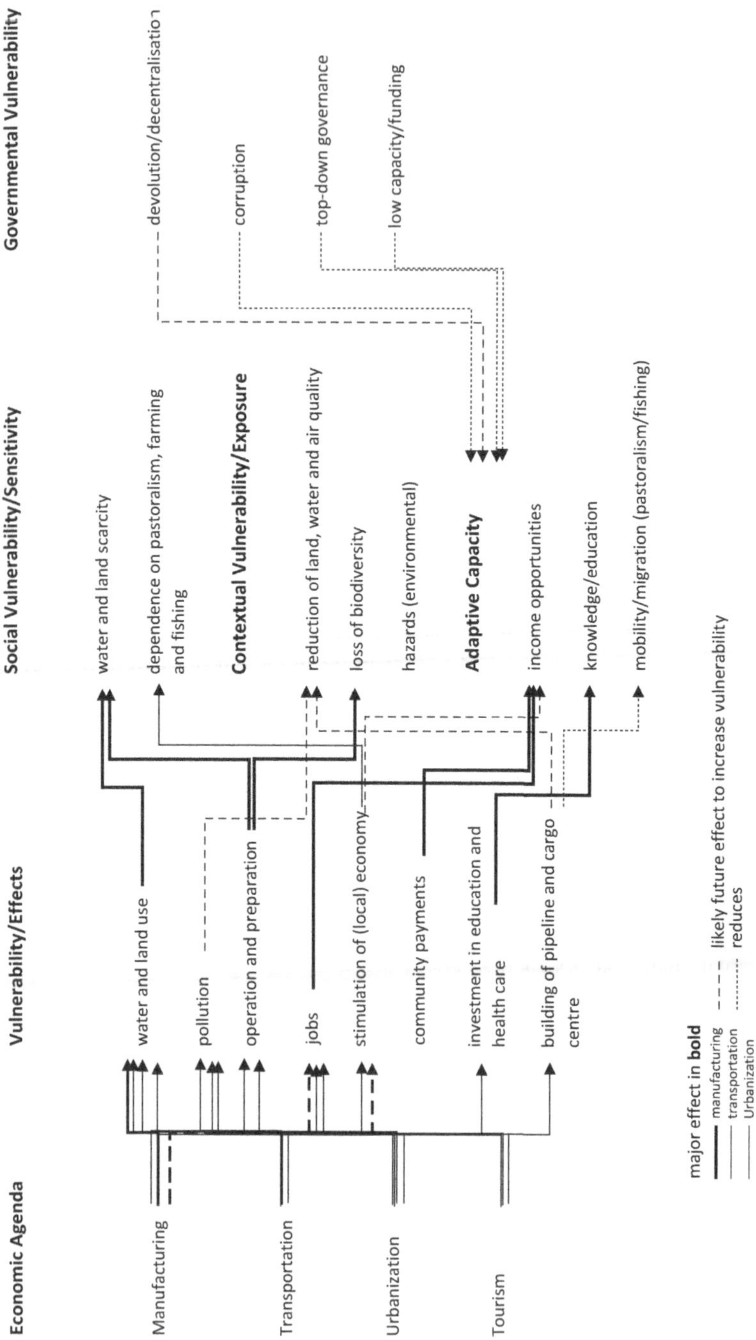

FIGURE 4.3 Effects of the Economic Agenda on the Vulnerability and Adaptability of Local Resource Users

(*Source:* The Author 2020)

horticulture, energy sites and the construction of a cargo centre. The main industry in Naivasha is agriculture, especially horticulture, making Naivasha the centre of a multibillion-dollar flower industry in Kenya[10] (Nyangena and te Velde 2012). Around Lake Wamala, the government allocated land to national and international companies to increase the share of rice growing, minerals and oil explorations to the country's GDP.[11] The main industry around Lake Wamala is fishing. However, rice growing and the exploration of minerals is constantly increasing. As mentioned in Chapter 3, Lake Naivasha is advertised as a tourist destination for both national and international visitors. Accordingly, the government promoted the set-up of hotel sites along the lake's shorelines.

Independent from the macro-economic projects, the majority of the people living at Lake Naivasha and Lake Wamala still depend on agriculture, fishing and pastoralism to purse a living (see Table 4.1). Hence, the higher the water and land shortages and dependence on those resources, the higher the sensitivity of the stakeholders. Land (including pasture) is already a scarce resource at Lake Naivasha. While the water levels of Lake Naivasha are sufficient, local resource users report serious water shortages because of reduced access routes to the lake's landing sites (Kapila 2018; Community Members Karagita Landing Site 2018; Community Members Central Landing Site 2019). Land and water around Lake Wamala is fairly available despite surface runoff due to rapidly changing and erratic weather patterns. However, limited access routes to the lake and an inadequate transportation network resulted in land and water shortages for the local communities (Community Members Lubajja and Lusaalira 2018, 2019; Kitzio 2019). Nowadays, through increasing deforestation, agricultural and grazing land should become more available; however, community members report on enclosed lands and declines in wetland and water levels.

Macro-economic and infrastructural projects aggravate water scarcity, as significant amounts of water are needed in every step of activities like flower growing or energy production, including drilling, watering and bathing of flowers, completion and fracturing of energy sites, and cooling. Studies show, for instance, that 7 to 13 litres of water is needed to harvest one flower[12] (Mekonnen et al. 2012). The Kenyan national government aims to increase the share of flower exports to the country's GDP.[13] Therefore, water is pumped from the lake into the flower farms directly, resulting in the decline of the lake's water levels. Additionally, the farms use chemicals and pesticides to boost the production of flowers. Inefficient and non-existent waste regulations resulted in the release of the used chemicals into the lake directly. The use of the aquifers to operate the geothermal sites could become a potential game changer to limit the pressure on Lake Naivasha's surface water. Studies show that if an aquifer is used to exchange heat with the earth, the used water returns to the same aquifer. Thus, when geothermal systems are operated commercially, they can eliminate millions of gallons of water (National Geographic 2013). However, Lake Naivasha faces challenges to implement its full potential. The lack of funding, technical expertise and corruption complicate and delay such projects (Vivekananda 2015; Kapila 2018).

Around Lake Wamala, infrastructural projects increased at the beginning of 2010. Beforehand, long-term poor agricultural methods, uncontrolled bush burning to increase the available land for agriculture, overgrazing and brick making put stress on the land and water resources (Kamese 2018; Kimbowa 2018). Given that Mityana's municipal waste collection does not include the fishing villages around the lake, the local actors burn or discharge their domestic waste into the lake. Similarly, these activities resulted in water quality deteriorations, a decline in fish stock and water-borne diseases. Industrial projects accelerated soil formation, wetland destructions and land and water shortages. The setting up of rice-growing fields and mining sites as well as deforestation resulted in the closing of fishing villages and the relocation of the people in the close surroundings of the lake.

Since the beginning of 2010, the impacts of the aforementioned behaviour were felt because the water quality of both lakes deteriorated significantly. Respondents reported the use of fertilizers, pesticides and other chemicals to increase the productivity rates in rice growing, agriculture and horticulture. Up to now, community members highlighted that the release of used chemicals and fertilisers in the lakes affect both the water quality and the fish stock (Community Members Central Landing Site 2018). Besides, villagers reported serious health problems after they consumed the water (Community Members Karagita Landing Site 2018; Peter 2018; Muigai 2019). This resulted in income reductions and serious health constraints on the communities. Therefore, the flower farms and Naivasha Water and Sewage Cooperation improved the communal access to clean water. They installed water tanks along the main road to fetch clean water. Community members and villagers reported that these tanks are filled by Naivasha Water with water trucks. Whereas initially the tanks were loaded up daily, interviewees reported that the tanks are filled less often nowadays. The large distances between the communities and Lake Wamala led to the drilling of boreholes and the construction of pumping stations. However, the process of providing an area-wide coverage has been slow. Some community members argued that neither the local nor the national government maintains the pumps or repairs them in case they break down. Therefore, this effort had a limited effect on providing sufficient amounts of water to the local communities.

Studies highlight that most pollution in the lake is not from fertilisers from the flower farms, but from soaps and detergents used by the village people in washing as well as untreated effluent from villages flowing into the lake (Harper et al. 2002). Further analysis showed that small-holder farmers and pastoralists push into the catchment areas[14] for agriculture and cattle grazing nowadays. More so, their agricultural techniques have become more aggressive as they use pesticides to boost their farming productivity. Accordingly, they contribute to soil degradations and the reduction of the lake's water quality. Observations and further studies showed that the small-holder farmers in the upper catchment of Malewa river and downstream villagers are also responsible for the lake pollution due to nutrient load, the wiping of oil and polluted cars and motorbikes or the release of waste into the lake (Harper et al. 2002; Mekonnen et al. 2012; Bokokamau 2018). In future, the issues of nitrogen, saline, mercury or phosphorus are likely to become more prominent, reducing the quality and quantity of rice and other agricultural products.

The Kenyan MTP I aimed to increase employment opportunities. Consequently, the labour-dependent horticulture and tourism industry attracts people from all parts of the country, including Nairobi. The majority of migrants come from Central, Western, Nyanza and Eastern provinces, and are predominantly from poor backgrounds. They come to Lake Naivasha in search of unskilled work in the flower farms and the hotels. On the other hand, the industrial activities around Lake Wamala do not attract unskilled workers from the country. The majority of the rural population moves towards Kampala in search of labour opportunities as Uganda Vision 2040 prioritizes non-agricultural job creation. As mentioned, the horticultural farms and hotel sites have the potential to offer direct and indirect employment opportunities to the local population and, therewith, might be able to stimulate the local economy. According to the respondents, Lake Naivasha has around 70 flower farms that employ over 150,000 people directly and over 500,000 indirectly (Vivekananda 2015; Peter 2018; Kimeta 2019). Whereas geothermal companies demand skilled workers, flower farms and hotel sites offer employment also to unskilled workers, for example, as flower pickers or waiters. To a certain extent, both could decrease the communities' dependence on agriculture or pastoralism. However, a former employee from one of the biggest flower farms indicated that the labour conditions and wages in the flower farms are low.[15] Employees do not receive mouth or respiratory passage protection. Resulting thereof, they are exposed to the chemicals and pesticides directly (Leipold and Morgante 2012; Peter 2018). Health constraints will likely be felt in the mid- to long-run, but remain subject to speculation at the moment. Lastly, female employees mentioned sexual abuse and gender-based violence by senior employees.

Consequently, the demand for affordable housing is increasing in the lake's basins. However, most of the labour migrants reside in one of the four informal settlements around the lake. In contrast, neither urbanization nor employment opportunities have revealed themselves for the people living around Lake Wamala. The population increase decreases the space in the four informal settlements around Lake Naivasha. Many communities around Lake Wamala live a cut-off life. Therefore, sub-qualified sanitation is also an acute issue that plagues the local daily life. In both areas, dumpsites and trash cans are very rare. Hence, they deal with the garbage and waste-water by dumping it on the streets and drainages. Thereupon, pollution and a disease-delivering environment is increasing.

Especially around Lake Naivasha, the investments in manufacturing and tourism resulted in road constructions to link the hotels, energy sites and horticultural farms to the main road. Community members at Lake Wamala did not report on any road constructions around the Lake. On the contrary, they highlighted that only the transportation network from Kampala to Mityana, and further to Fort Portal is improved. Workers from the flower farms and villagers from Karagita confirmed that the government improves neither the access routes to the four landing sites, nor the access routes to the informal settlements. Throughout the research stays at

Lake Wamala, one was able to observe the state of the access way from Mityana town to the remaining three landing sites. The nearest landing site, Lubajja, is located four kilometres away from Mityana town. Because of its soil conditions (sand), during the rainy season, motorbikes and cars are unable to pass. Fishermen from the landing sites around Lake Wamala highlighted that they were unable to sell fish and other agricultural products to merchants because trucks only pass by Mityana. During the dry season, additionally, they demonstrated that they often have to walk up to Mityana to sell their products, either at the local market or to merchants from Kampala, as trucks pass by their landing sites less frequently.

The previously mentioned oil explorations do not take place at Lake Naivasha or Lake Wamala. Nevertheless, in the future, there is a risk that the pipeline between the extraction site, the refinery and the port city of Lamu (Kenya)[16] and the city port in Tanga (Tanzania)[17] will disrupt migration routes of pastoralists as well as affect the cultivation of farmland. As a result, internal migration is increasing, pushing displaced people closer to labour-offering areas, like around Lake Naivasha, for example. As a consequence of the macro-economic projects, the flower farms and hotels were allowed to build their sites along the lake shores, thereby decreasing the available land for agriculture and cattle grazing. Furthermore, the government invested in transportation routes linking the industrial sites with Naivasha and the main road. The pressure on small-holder farmers and pastoralists increased because a higher amount of local resource users is in immediate need to use the remaining land resources (Vivekananda 2015).

To sum up, both lake basins experienced deforestation, wetland degradation and overexploitation. Furthermore, the pressure on the lake's resources is increasing due to population growth, aggressive agriculture and climate change. Lastly, poor resource management leads to water and soil shortages.

4.4.2 *Exposure*

Studies show that unsustainable land management increases the effects of climate change on ecosystems and societies (IPCC 2019). But the economic developments undertaken at Lake Naivasha and Lake Wamala, at least so far, did not increase the climate exposure for the stakeholders (see Table 4.1). However, the economic projects undertaken expose especially the local communities to environmental changes (i.e. resource access and resource distribution). These environmental changes, furthermore, interact with and accelerate changes in temperature and rainfall (e.g. heat waves or heavy rainfall). Community-based organizations have emphasized the long-known fact that wetland degradation, deforestation or poor waste management on water and land resources is affecting 'water supply and water resource management' (Uluocha and Okeke 2004) and also threatens the surface and groundwater quality (Konbusingye 2019). Studies conducted in countries like Nigeria, Niger, Malawi and South Africa demonstrate that economic infrastructural projects without environmental impact assessments can have detrimental effects on the environment and the local population, especially where resource governance or environmental regulations are lacking or not enforced (for Nigeria Uluocha and Okeke 2004; for Niger Adekola and Mitchell 2011; for Malawi and South Africa Schuyt 2005). Apart from the changes in resource access and resource distribution, Schuyt (2005) highlights the economic consequences[18] for the actors who are most dependent on land and water resources for their daily survival (i.e. local populations, farmers and pastoralists) (Schuyt 2005).

Additionally, water, land as well as air is polluted through the burning of waste, bushes and charcoal. This, in turn, increases the exposure of the local population to water-borne and other diseases like diarrhoea, cholera or typhoid fever (Otieno 2019; Yakub 2019). Respondents from Kenyan and Ugandan governmental agencies and county/district governments commented that the governments acknowledge that the countries are vulnerable to illegal dumping of waste and other substances (Mbaisi 2018; Tindimungaya 2018). Consequently, both governments drafted national policies for waste management regulations. Observations and interviewees confirmed that, however, the implementation of the policies is limited, especially in rural and slum areas. Solid waste management often remains subject to local or community-based initiatives (Kaka 2019; Otieno 2019).

Whilst stakeholders, who do not depend on agriculture, pastoralism or fishing to pursue a living, did not report on negative changes in their environment, community members and local resource users mentioned changes in their resource access, resource distribution and the economic consequences thereof. Small-holder farmers and residents of the informal settlements around Lake Naivasha stated that farmland becomes unavailable due to the set-up of flower farms and hotels or abrupt closing of access routes to farmland or public areas.[19] Correspondingly, small-holder farmers encroach into the lake's shorelines through one of the four remaining landing sites in search for farmland. Furthermore, they claimed that due to changes in water levels, the farmlands are flooded or the crops are either small or do not sprout because they use pesticides or polluted water from the lake. Concerning small-holder farmers at Lake Wamala, community members pointed out that small-holder farmers are hardly found at the lake anymore. Most land is used by macro-economic companies for rice growing, mineral activities or timber production and, thereupon, is gazetted nowadays. Consequently, an increase in farming activities in the wider surroundings of Lake Wamala was observed. As a result of the longer distances to the lake or the pumping stations, small-holder farmers rely substantially on rainfall to water their crops. However, unsustainable farming and economic activities accelerate climate change. This results in the unpredictability of seasonal changes and, therewith, farming becomes more difficult. Small-holder farmers stated that crops are washed away or dry up more frequently which leads to economic losses.

At Lake Naivasha, fishermen reported smaller fish stock as a result of overfishing. Small-holder farmers who do not have farmland anymore, or people who worked in flower farms often turn to fishing to sustain a living. Resulting thereof, the number of people who fish, either illegally or using unlicensed fishing gear, is increasing. In the case of Lake Wamala, as of 2021,[20] there are three remaining landing sites where fishermen can access the lake directly. Because of their relocation into the wider surroundings of Lake Wamala unlike Lake Naivasha, most fishermen turn to farming activities as they are unable to approach the lake's shorelines anymore. Fishermen at both lakes reported that the release of chemicals, pesticides and other substances led to a reduction in fish sizes and stock. However, to today's knowledge, there does not exist a study on the effects of the water quality on Lake Naivasha and Lake Wamala's fish stock. Therefore, in the future, this issue might become more prominent, and remains subject to speculation for the time being.

Pastoralists, especially Maasai and herders from Laikipia county, claimed that the reduction of available land for grazing pushes the cattle closer to the road. The noise from the cars, motorbikes or trucks scares the livestock on the one hand, and on the other hand cattle are injured or killed in road accidents more often. Community members agreed on increased traffic because of jeeps and trucks carrying heavy machinery related to the building of the geothermal sites or the cargo centre. Lastly, residents from the informal settlements highlighted a decrease in water access, as the amount of people pushing onto the remaining landing sites and the few water access points in the informal settlements is increasing. Furthermore, community members around Lake Wamala and villagers at Lake Naivasha emphasized long walking distances to fetch water from the pumping stations or the lake for daily activities. Moreover, villagers from the informal settlements reported on unacceptable living conditions, given the increasing number of residents. The unavailability of waste collection, basic sanitation and energy increases air, land and water pollution, and the outbreak of diseases results in higher levels of insecurity.

4.4.3 Adaptive Capacity

Generally, adaptive capacity is mainly determined by the knowledge, skills, resources and options the different stakeholders have to cope with external changes (IPCC 2014). Key adaptation strategies include income opportunities, increase in education and knowledge, the effectiveness of political institutions or migration.[21] Pastoralists mentioned and studies confirmed that internal migration was and is always a key adaptation strategy to counter lack of water and pasture (Schilling et al. 2014; Schilling et al. 2015; Balama 2019).

Especially the small-holder farmers and the people from Eastern and Western Kenya migrate to Lake Naivasha in search of work in the flower farms and hotels (Kapila 2018; Peter 2018). Additionally, pastoralists and small-holder farmers claimed that economic sites and the construction of above-ground pipelines had altered migration routes and blocked farmland. A minority of the interviewees argued that currently ongoing road and railway constructions as a part of the national economic development plans are likely to affect migration routes and farming activities. However, observations have shown that the effects of above-ground pipelines are limited. The researcher considers the future constraints of the railway and road constructions more severe.

Both national development agendas stress the importance of employment and the stimulation of the (local) economy related to the undertaken infrastructural investments. Whilst some Ugandan interviewees reported that economic companies offer payments to the communities to compensate them for the loss of their land (Muhindo 2019), others argued that local communities are not compensated financially, but rather through the provision of additional land (Twesigye 2019). When asked about community payments at Lake Naivasha, the participants did not mention any compensation, neither financially nor through the provision of additional land. Generally, the availability of financial resources increases the adaptive capacity of community members, at least in the short term. However, this also has other implications. The county and district representatives described the geothermal companies, the upcoming cargo centre and other economic projects as promising for income opportunities. However, the low level of education, especially in these rural areas, is a disadvantage for the local communities as they cannot compete with others for economic and employment opportunities. There have been some rare attempts by the governments to increase the level of education. Along the access roads to Lake Wamala and in the informal settlements, a few classrooms have been built for primary teaching. However, these schools have not begun to operate as equipment (i.e. chairs or tables) either were stolen by the local communities or had not been provided in the first place. Added to that, neither have teachers been ordered to these schools, nor have announcements been placed to attract teachers to the schools respectively. Accordingly, local resource users are often left to work in flower farms, hotels or as artisanal miners. The resilience to these changes is worsened by the strong population growth[22] which is highly likely to further intensify resource shortages and to exaggerate the poverty levels.

These findings suggest that the adaptive capacity of local resource users is low, because the state of their overall personal living conditions can be described as being multi-dimensionally poor. According to the World Bank Group, this category includes not only the criteria of income and consumption but also ‘other aspects of life [which] are critical for well-being, including education, access to basic utilities, health care, and security’ (World Bank Group 2018 5). Uganda has both a lower per capita income and a higher Gini index and thus, compared to Kenya, a more unequal distribution of income (CIA 2020; see Table 4.3). Similarly, the level of human development is higher in Kenya than in Uganda (UNDP 2021a). Moreover, the under-five child mortality rate is lower in Kenya than in Uganda (World Bank Group 2020h). However, both countries made significant progress increasing their health service delivery.[23] Uganda, especially, increased its percentage of rural people with access to improved drinking water sources (AQUASTAT 2020). Data only shows how the number of access points to drinking water changes, but does not include information about the condition or kind of access points. Hence, it cannot be concluded from these findings if water is actually provided through the pumping stations or the water access points.

In terms of education, both countries have improved. Nevertheless, there is still a gap between the rural and urban mean years of schooling. In rural areas, children

TABLE 4.3 Determinants of Adaptive Capacity to Economic and Environmental Changes

	Determinants of Adaptive Capacity								
	Economic Resources				*Health and Water*		*Education*		*Institutions*
	Gross domestic product per capita in PPP USD (2021)	*Gini Index (year estimate)*	*Human Development Index (2021)*	*% of people below poverty line (2017)*	*Under-five mortality rate per 1,000 live births (2019)*	*% of rural population using improved drinking-water sources (2017)*	*Education Index (2019)*	*Global Knowledge Index (2020)*	*Corruption Perceptions Index (2020)*
Kenya	4.220	40.8 (2021)	0.601	36.1	43.2	56.8	0.534	56.2	31
Uganda	2.400	42.8 (2021)	0.544	21.4	45.8	75.8	0.523	32.3	27

(*Source*: The Author 2020, based on CIA 2020; UNDP 2019, 2021a, 2021b; Transparency International 2018).

PPP USD = Purchasing Power Parity in USD. Gini Index (Range 0–100) = the degree of inequality in the distribution of family income in a country (World Bank Group 2021). Human Development Index (Range 0–1) = measuring life expectancy, education and income (UNDP 2021a). Education Index (Range 0–1) = mean years of schooling and expected years of schooling (UNDP 2021b). Poverty line = percentage of the population living below 1.9 USD per day (CIA 2020). Global Knowledge index (Range 0–100) = measuring education, research, development and innovation, information and communications technology and economy (UNDP 2021a). For all mentioned indices, higher values indicate higher level of development. For Gini Index, higher values indicate higher levels of inequality. Corruption Perceptions Index (Range 0–100) = perceived levels of public sector corruption according to experts and businesspeople (Transparency International 2020). Higher values indicate less corruption transparency.

are often unable to attend school since either the walking distance to the nearest school is too far or the way too dangerous, or the children have to assist their parents in daily work activities at home (i.e. farming or fishing, or cleaning and cooking). The long school closings (two years in Uganda and one year in Kenya) have, however, setback the slowly achieved processes in attendance rates. In some cases, both factors prevent children to even attend primary school in rural areas. The percentage of people living below the poverty line is lower in Uganda compared to Kenya (CIA 2020). For Uganda, the lower levels of income and, therewith, higher levels of poverty cannot only be explained by using the data from the Gini index and GDP per capita PPP. There seem to be two possible explanations for this: (1) in Kenya, the number of people living in urban areas[24] is higher than in Uganda. Given that living expenses are greater in urban areas compared to rural ones, there are more people living below the poverty line. (2) Another possible explanation for this is that in Uganda, local resource users often cultivate what they need to sustain for a living and, therefore, do not consider themselves as poor. Taking all indicators into account, Kenya has a higher adaptive capacity. Uganda's adaptive capacity is lower regarding institutional determinants. However, both countries differ by only one point regarding their level of corruption. Hence, both find themselves among the lower third of the most corrupt countries in the world (Transparency International 2018). Resulting thereof, corruption and political kinship limit the enforcement of the institutional framework, restrict the implementation of economic and social reforms which could increase the adaptive capacity of local resource users.

As indicated, adaptive capacity is not only determined by socio-economic factors. For the successful implementation of adaptation processes, the political system's institutions and organizations clearly need to be considered to get a comprehensive understanding of adaptive capacity as well. Institutions endure and persist throughout the adaptation process and shall support individuals and communities in times of adjustment and change (Nelson et al. 2007).

4.4.4 Governmental Vulnerability

All of the discussed economic and social indicators of vulnerability, exposure and adaptive capacity are enclosed by the influence and effectiveness of political institutions. Economic infrastructure, poverty and income, access to and the distribution of resources within a population are all institutionally determined, hence central to the overall analysis of vulnerability, and possible resulting conflict dynamics thereof.

Devolution in Kenya is completed, and therefore, it is expected that political, economic and social services should come closer to the people. This will likely affect the level of adaptive capacity of local resource users in the mid to long term. On the other hand, Uganda is still in the process of decentralising its institutional framework. The key aim of both processes is to bring services closer to the people and to reduce the imbalances between the different regions and counties within Kenya and Uganda (Vasquez 2013). However, both county and sub-county officers were unanimous in the view that they lack the capacity, knowledge and resources to effectively deliver services to the people. In Uganda, the district and catchment management officers highlighted that there are no financial resources available to both maintain the catchment committees as well as to set up institutional structures, including office premises or to hire employees. The Lake Albert Catchment Manager claimed that there does not exist any legal backing from national authorities, thus the decentralized structures are still in their infants (Kitamirike 2019).

The principal idea of devolution and the introduction of decentralized structures incorporate the sharing of governmental functions, power and resources between the national and county/district level (Constitution of Uganda 1995; Constitution of Kenya 2010). As a result, a fund was established under the Ministry of Finance and the National Treasury to give both less developed counties and districts, as well as their communities the opportunity for empowerment. The aim is to enable them to 'catch up' with more developed regions and communities (Ibid.). On the basis of the various indicators shown earlier, it is clear that urban areas and economic stakeholders, whose revenues from horticulture, tourism, energy or mining contribute to an increasing GDP, are the most favourable by national decision-makers, while mostly unskilled and local resource users, who are dependent on agricultural, fishing and pastoralism, appear to be the most marginalized. There is strong evidence that national-level authorities provide statements of political decision-making to favour policies, stakeholders and regions serving their economic development plans.

The named challenges have undermined the effectiveness of devolution and, therewith, the financial fund cannot serve its original objectives. One major challenge is that ongoing political decision-makings work against the principle of separation of powers and still follow a top-down approach. Accordingly, a number of resources and water related projects are not related to the needs of the community and, hence, are not realized. Another challenge includes the low capacity of the sub-county and district authorities to realise projects related to the needs of the communities. Often, the county and district officers are not given a real opportunity to make decisions on their own, since the national leaders make all the important decisions and expect the communities to accept them. Thus, they do not have the autonomy over the decision-making process, although it is guaranteed by the existing laws. More so, community members and the rural population are often unaware of the decentralized structures and, therefore, hardly participate in critical stages or question the decision-making processes. Another reason for this inactivity is that community members fear arrests, the loss of jobs or access to resources[25] if they participate in community meetings or challenge sub-county officers. Most of this information was retrieved from focus-group discussions. However, there was not an opportunity to cross-check this information with other sources. Accordingly, a note of caution is advisable. Another challenge which should not be underestimated is corruption. During the drafting process of devolution and the decentralized system, critical voices argued that the management of the new system should not be in the same hands as of the people who established it (Wamuyu 2016). It is worth noticing that it has been the case in both countries that sub-national institutions followed national powers and guidelines as often relatives of national authorities hold power on a sub-national or local level of decision-making. Moreover, some community-based organizations follow the top-down approach as they get access to groundwater, land or other public services in return.

The present discussion revealed that the implementation of the governmental structures according to the laws in place stagnated. When combined, the contradiction between the economic development plans, the laws in place and the institutional effectiveness indeed have the potential to increase existing inequalities which both countries have grappled with since colonial times. It is generally accepted by both Kenyans and Ugandans that what is lacking is political will to make the devolved and decentralized system of government work. It seems that many powerful people (at a political and economic levels) have vested interests in centralized power structures again. Resulting thereof, they are responsible for the high levels of vulnerability and low levels of adaptive capacity associated with environmental and economic changes for the local resource users around Lake Naivasha and Lake Wamala. This, finally, increases the inequalities in the regions and results in higher insecurity perceptions.

4.5 Vulnerability and Conflict Risk

This section discusses the conflict risks which are associated with the changes in resource access and resource distribution. It firstly looks at the social implications the resource users face. The implications are actor- and location-specific because they depend on the stakeholders' vulnerability to the mentioned changes. The section moves on to discuss the likelihood that conflicts erupt between and within the stakeholder groups. In particular, the dimension of community-community conflict is focused on.

4.5.1 Social Implications

The reduction of water, land and air quality will negatively affect food security and the daily livelihood of especially rain-fed dependent stakeholders around Lake Naivasha and Lake Wamala. The resilience of economic stakeholders, county and district officials and senior employees from non-governmental and community-based organizations is higher because they have, on the one hand, enough funds available to buy land titles or water to sustain a living. On the other hand, their level of education is greater as they have been able to attend primary and secondary school, or even vocational training or university and, thus, often move to Nairobi in search of employment in the industrial sector.[26] Moreover, political connections and ethnic affiliations grant those actors preferential access to higher education, springs[27] and land titles[28] and, hence, reduce their dependency on the lake's water and public land to sustain a daily living.

Because of environmental degradations, biodiversity is lost, which, in turn, accelerates the effects of climate change in the two basins and makes sowing unpredictable. Based on the IPCC scenario, the impacts of climate change are likely to result in some 1.8 million extra cattle which could be lost till 2030 because of increased drought frequency. It is projected that the value of the lost animals and production foregone amount to 630 million USD (Herrero et al. 2010; IPCC 2019). Likewise, the IPCC AR 5 scenario forecasts compound stress on water resources, reduced crop productivity and livelihood and food insecurity as well as higher risks of water-borne diseases due to increased temperatures, heat waves, droughts and more intense precipitations (IPCC AR 5 2014). In addition to the losses caused by unpredictable seasonal changes, the lake level fluctuations at Lake Naivasha led to agricultural losses, as soil and seeds have been washed away. Due to the limited space around Lake Naivasha, small-holder farmers move into the catchment areas at one of the four remaining landing sites to do farming. The decline in water and grazing areas forces pastoralists to move farther away from their original migration routes in search of water and land. Hence, the probability that small-holder farmers and pastoralists share the space and available resources at the lake's access points more frequently increases. Whereas pastoralists in Uganda have not yet moved to central Uganda, nowadays pastoralists are found around Lake Naivasha.

Around Lake Naivasha, fishermen have united in fishing associations to increase their level of income. Most associations sell their fish directly to traders who, in turn, sell the fish in the market in Naivasha or Nairobi. However, the number of fishermen is increasing, as former employees of the flower farms, unemployed youth from the informal settlements or erstwhile farmers turn to fishing. This increases the competitiveness between the unions and results in overfishing and illegal fishing methods. The most obvious finding to emerge from this analysis is that the different groups of local resource users push for the remaining four landing sites nowadays, and therefore, social tensions are increasing around Lake Naivasha. On the contrary, the pressure on farmland is increasing in the surroundings of Lake Wamala, as former fishermen are forced to change their profession because fishing is not possible anymore.

As a result, local resource users already experience reductions in crop and fish stock. Moreover, pastoralists reported on starving cattle as well as the loss of cattle due to road accidents. Thus, most resource users live below the poverty line and are unable to purchase water or food from the nearby towns. Likewise, the lack of adequate funding makes it impossible for parents to pay school fees or purchase school materials for their children. This might result in higher unemployment rates in the mid to long term. The lack of sufficient financial resources and a social welfare system leads also to the continuously high birth rate (see Table 4.3). Another important social implication arising from internal migration and high birth rates are increasing social tensions between the villagers in the informal settlements over housing, the few water stations and the employment opportunities for unskilled workers.

4.5.2 Conflict Risk

The information from the interviews have been used to calculate the risk over the outbreak of water-related conflicts using the software MAXQDA and Excel (see Chapter 1). The section *Risk Conflict Outbreak* assesses the general risk over the outbreak of water-related conflicts. The other five sections assess the likelihood that water-related conflicts erupt between one stakeholder group and another. The social implications which result from the economic development agendas is higher for the local resource users compared to the others. The current study found out that the local resource users at both lakes face significant challenges which result from political, socio-economic and environmental changes. Based on the data evaluation, the risk over the eruption of water-related conflicts is the highest within the group of the local-level actors (49 per cent) (see Figure 4.4). The figure shows that the group of local resource users has, by far, the highest values of conflict risks, not just within its stakeholder group but also across the groups. As previously indicated, the conflicts between local resource users and national (21 per cent) or international actors (16 per cent) are non-violent and are, at most, demonstrations or the expression of critical voices by community leaders against the economic projects. A closer inspection of the figure shows that the conflict risk between local-level actors and sub-county- (5 per cent) or county-level actors (8 per cent) is lower compared to national and international ones. A possible explanation for these results may be the lack of adequate knowledge about the governance competencies of sub-national actors. Another possible explanation for this is that neither devolution nor decentralization is implemented sufficiently.

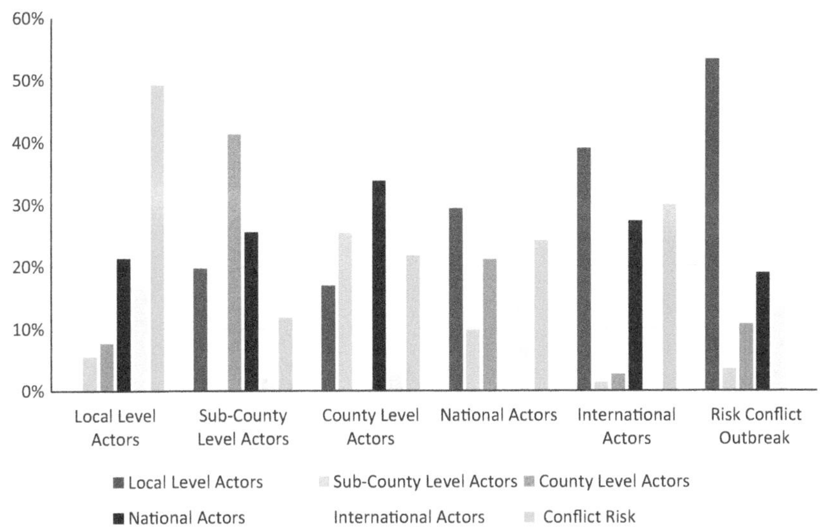

FIGURE 4.4 Percentage of Identified Conflict Risk Within the Stakeholder Group and Across the Groups in Kenya and Uganda

(*Source:* The Author 2020)

The previous reason can also be attributed to the low-conflict risk between sub-national- and national-level actors. Furthermore, the analysis showed that kinship ties also limit a stricter enforcement of devolution and the decentralized structures. Even though national and international actors reported a clash of interest the economic orientation, the risk of the outbreak of a conflict can be weighed as being close to zero. This positive relationship is explained by, firstly, the high level of dependency of the national actors from the financial development contributions on mostly international organizations. Secondly, national actors benefit from the undertaken economic investments of international economic actors, hence the government of both countries are reluctant to express criticism regarding the international actors' behaviour. More so, governmental actors give international and national economic actors plenty of rope as they are afraid that economic actors close their sites at one of the lakes to set up the same business in another country. The latter happened already around Lake Naivasha. Some flower enterprises shut down and moved to Ethiopia as the Government of Ethiopia is offering better economic conditions[29] than Kenya (Kimeta 2018).

The discussion and data evaluation noted a strong relationship between local resource users' high sensitivity to external changes, low adaptive capacity and higher levels of insecurity. These results, further, support the perception of higher conflict risks within the group of local-level actors compared to all other stakeholder groups. The data in Figure 4.5 highlight that local-level actors are very

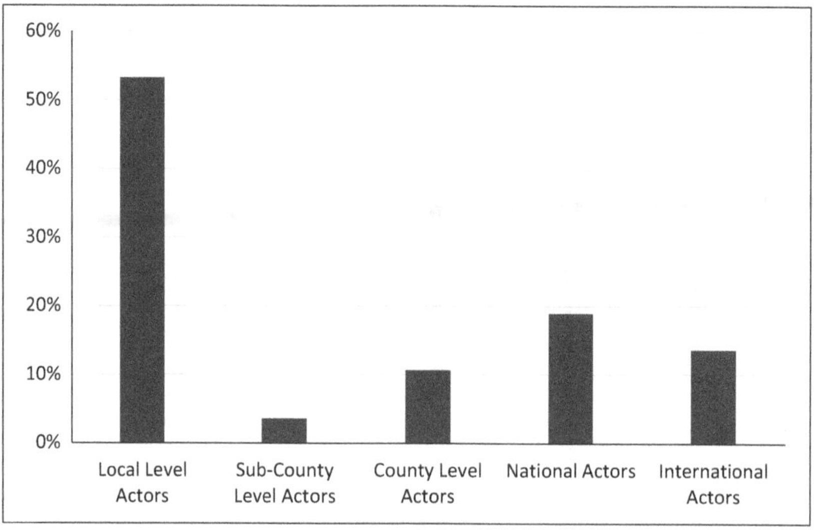

FIGURE 4.5 Percentage of Likelihood for Outbreak of Conflicts Within the Stakeholder Categories

(*Source:* The Author 2020)

vulnerable to conflict (53 per cent) in Kenya and Uganda. International (14 per cent) and national actors (19 per cent) reported that clashes of interests are possible conflict drivers. However, they mentioned that those can be solved through power sharing or grant benefits. Local-level actors indicated that their social tensions are settled by often violent behaviour (see Chapter 6).

4.6 Brief Summary

This chapter aimed to address two questions: Firstly, why do conflicts only erupt between local resource users? Secondly, to what extent does climate change cause water shortages at both lakes? All stakeholder groups around Lake Naivasha and Lake Wamala face significant challenges in resource access and resource distribution which originate from political and economic changes. The implementation of the economic agendas is moving forward. The current extractions, infrastructural projects or horticultural activities promote employment opportunities for unskilled workers and increase the economic and infrastructural development at both lake basins. Accordingly, this might reduce the vulnerability of especially the local resource users because the economic developments shall increase their adaptive capacity. However, the economic projects accelerate environmental degradations, put pressure on the water levels and increase the pollution of soil and water. Therefore, the economic projects do not strengthen the local people's adaptive capacity. Instead, the study showed that these economic developments are likely to outstrip their vulnerability further.

The majority of both Lake Naivasha and Lake Wamala's people are strongly dependent on agriculture, pastoralism and fishing for their income and employment. Consequently, the industrial projects are expected to undermine their food security, resilience, resource access and, therewith, their overall livelihood perspective in the region. Their low resilience is significantly undermined by the strong population growth which will increase the demand for food, water and income opportunities in the hotels, flower farms or artisanal mining sites. Both governments have so far failed to provide local resource users with basic services, including education, water infrastructure, health centres and security. Added to that, environmental changes aggravate climate-related effects on the availability of resources. Therefore, the gap between the supply and demand of critical resources (i.e. water and land) is widening. Nonetheless, an obvious finding to emerge from the analysis is that there is not sufficient evidence to draw a direct link that climate change increases water shortages in the two study areas.

It may be the case that collective vulnerability is intensified in the mid to long term by "exogenous" environmental changes which will occur due to global climate change and environmental degradation. This part of the study also accords with earlier observations, which showed that the links between climate change, resource availability and conflict are often indirect and complex (see, e.g. Buhaug 2015; Schleussner et al. 2016; Adams et al. 2018). Nevertheless, the two aspects of vulnerability are obviously interlinked. At the community level, social vulnerability

TABLE 4.4 Summary Individual and Collective Vulnerability and Adaptive Capacity to Changes for Both Kenya and Uganda

	Indicators of Vulnerability	*Adaptive Capacity in Relation to Changes*
Individual Vulnerability	Distribution of Incomes, Diversity of Economic Assets, Education, Dependency on Resources	Poverty, Access to and Distribution of Resources
Collective Vulnerability	GDP, Inequality, Development Levels, Indicators of Institutional Arrangements, Formal and informal security	Levels of Infrastructure and Development, Institutional Effectiveness

(*Source*: The Author 2020)

is affected by the relative distribution of income, access to and diversity of economic assets, access to and the distribution of resources and by the operation of informal social security arrangements. Further, individual vulnerability to economic and environmental changes is determined by the institutional arrangements which organize warning, planning and other services summarized as "collective" vulnerability in Table 4.4. The likely future extent of the social implications depends on firstly, how ongoing and planned macro-economic operations reduce the available land further Secondly, the creation of better employment and education opportunities can limit the dependence on pastoralism, farming or fishing.

Together, these results provide important insights into the interrelation between high levels of sensitivity and low levels of adaptive capacity for especially local resource users compared to the other stakeholder groups. These results suggest that there is an association between the levels of conflict risks, the levels of sensitivity and adaptive capacities which result from the environmental and economic changes. The results in this chapter indicate that the risk of the outbreak of conflicts is significant for the local-level actors regarding water access and water distribution. The next chapter, therefore, moves on to discuss and to analyse the conflict behaviour and dynamics between the local-level actors.

Notes

1 As there does not exist a climate station at Lake Wamala directly, the climate data for Mityana serves as a reference.

2 The Foundations for the Pillars include disaster risk reduction and ending drought emergencies. While disaster risk reduction is an environmental risk and can lead to an increase of people's vulnerability, respondents did not mention this aspect in the case of Lake Naivasha. Therefore, it is not assessed in detail.

3 In 2008, the ratio of investment stood at 19.61 per cent. In 2012, it stood at 21.48 per cent (World Bank Group 2020d).

4 In 2018, Kenya's GDP growth per capita stood at 3.9 per cent (World Bank Group 2020e).
5 The manufacturing share of GDP is constantly decreasing from its all-time high in 2007 (12.79 per cent).
6 Information and Communication Technology.
7 In 2018, the growth rate speeded up compared to the previous years (4 per cent in 2017 and 4.6 per cent in 2016) and stood at 6.1 per cent in 2018 (World Bank Group 2020g).
8 In 2018, the GDP per capita income slowed down to 642.777 USD, compared to 739.374 USD in 2015 (World Bank Group 2021).
9 Up to date, drafts of the NDP III exist only. Therefore, the mentioned strategic priorities are preliminary only and are subject to change.
10 Cut flowers from Lake Naivasha account for more than 70 per cent of the country's cut flower exports. Cut flowers generate 11 per cent of Kenya's total foreign exchange revenue (roughly 830 million USD) in 2012 (Nyangena and te Velde 2012).
11 According to a study carried out by UKAID et al. (2018), the extraction of minerals is estimated to generate a value of 500 million USD (roughly 3.5 per cent of the country's GDP) in 2010 (UKAID et al. 2018).
12 Compared to one flower, a villager in one of the four informal settlements has 3 to 5 litres of fresh water available on average a day.
13 In 2017, the flower industry contributed 1.3 per cent to the country's GDP (Deutsche Welle 2017).
14 According to the law, catchments are protected areas located at the lake's shorelines directly. According to the law, catchments cover 50 to 100 metres.
15 On average, a harvester earns between 50 and 100 USD per month (Vivekananda 2015; BBC 2017).
16 Tullow-Total oil pipeline, linking the oil fields in the Lockichar basin with the city port of Lamu, launched in August 2019.
17 Uganda-Tanzania Crude Oil Pipeline, linking the oil fields in Hoima with the city port of Tanga in Tanzania, launched in 2015.
18 For example, poverty, income reduction and low levels of education.
19 Due to unclear land tenure, small-holder farmers were used to access free land to cultivate crops. However, in some cases, the used land was either private or sold by the government to individuals who shut off the land without prior notice.
20 During the field visit in summer 2018, 2019 and early 2022, the researcher was able to access four landing sites. One year later, the landing site Katigo did not exist anymore.
21 The determinants for adaptive capacity are chosen because they are all available for the two countries studied at the same time. Furthermore, these variables are recognized by international agencies and organisations. Lastly, the determinants are used across all disciplines as accepted variables.
22 According to UNDESA, the total fertility (children per women) stands at 3.26 in Kenya and 4.44 in Uganda (UNDESA 2020).
23 Health service delivery increased in urban areas significantly whilst access to hospitals or clinics is limited in rural areas.
24 The number of people living in urban areas does not coincide with the number of people being employed.
25 Community members at Lake Naivasha reported that the Cargo Centre at Lake Naivasha will limit the forest, land and water coverage as it is planned to divert water from Malewa river to the Cargo Centre directly. However, one can hardly find critical voices or even community mobilizations against this infrastructural project. Local resource users and a member of a community-based organization mentioned the disappearance of colleagues and friends after they have voiced their concerns about the economic project.
26 Especially young people move to urban areas (Nairobi, Mombasa or Kampala) in search for education and employment. Studies have shown an increase in unemployment rates and insecurity especially in slum areas in Nairobi and Mombasa as a result of the

mentioned internal migrations (see, e.g. Shem Otoi 2016; Peter et al. 2018; Otieno 2019). However, it is beyond the scope of this study to discuss social implications and resulting conflict dynamics thereof.

27 Around Lake Naivasha, it was reported that senior employees from non-governmental and community-based organizations have been granted water access through springs or water tanks at their private premises. It was indicated that these community and non-governmental organizations decrease their level of criticism regarding the economic projects undertaken and resulting land and water shortages and resource access points in return. The researcher was able to observe this once. However, these data must be interpreted with caution because it is not based on hard evidence or multiple observations and statements.

28 Around Lake Wamala, respondents reported that kinship, political connections or rich businessmen acquired public land through the national government to sell it to international companies shortly after.

29 For example, tax reduction for a period of over 20 years.

Bibliography

Adams, C., Ide, T., Barnett, J. and Detges, A. (2018): Sampling bias in climate-conflict research. *Nature Climate Change* 8, 200–203.

Adekola, O. and Mitchell, G. (2011): The Niger Delta wetlands: Threats to ecosystem services, their importance to dependent communities and potential management measures. *International Journal of Biodiversity Science, Ecosystem Services & Management* 7 (1), 50–68.

Adger, W. N. (1999): Social vulnerability to climate change and extremes in coastal Vietnam. *World Development* 27 (2), 249–269.

Adger, W. N. (2006): Vulnerability. *Global Environmental Change* 16 (3), 268–281.

Aquastat (2020): *Data Query*. At: www.fao.org/nr/water/aquastat/data/query/results.html [retrieved last 8 February 2020].

Balama, J. (2019, 15 August): Interview by Renner, J. Nairobi, Kenya.

BBC (2017): *No Bed of Roses: The Kenyan Flower Pickers Fighting Sexual Harassment*. At: www.bbc.com/news/business-39103419 [retrieved last 6 February 2020].

Blaikie, P., Davies, I., Cannon, T. and Witner, B. (1994): *At Risk: Natural Hazards, People's Vulnerability and Disasters*. Routledge. London.

Bokokamau (2018, 9 August): Interview by Renner, J. Naivasha, Kenya.

Buhaug, H. (2015): Climate-conflict research: Some reflections on the way forward. *WIREs Climate Change* 6, 269–275.

CIA (2020): *The World Factbook. Kenya*. At: https://www.cia.gov/library/publications/the-world-factbook/geos/ke.html [retrieved 10 February 2020].

CIA (2022a): *The World Factbook. Kenya*. At: www.cia.gov/library/publications/the-world-factbook/geos/ke.html [retrieved last 10 February 2022].

CIA (2022b): *The World Factbook. Uganda*. At: www.cia.gov/library/publications/the-world-factbook/geos/ug.html [retrieved last 10 February 2022].

Community Member Karagita Landing Site (2019, 20 August): Focus Group Discussion by Renner, J. Karagita Landing Site, Kenya.

Community Members Central Landing Site (2019, 20 August): Interview by Renner, J. Naivasha, Kenya.

Community Members Karagita Landing Site (2018, 11 August): Interview by Renner, J. Karagita Landing Site, Kenya.

Community Members Lubbajja and Lusaalira (2018, 24 August): Interview by Renner, J. Lubbajja, Lusaalira, Uganda.

Community Members Lubbajja and Lusaalira (2019, 25 July 2019): Interview by Renner, J. Lubbajja, Lusaalira, Uganda.
Deutsche Welle (2017): *Sind Rosen für Valentinstag schlecht für die Umwelt?* At: www.dw.com/de/sind-rosen-f%C3%BCr-valentinstag-schlecht-f%C3%BCr-die-umwelt/a-37541851 [retrieved last 9 February 2020].
Food and Agricultural Organisation (2018): *National Gender Profile of Agriculture and Rural Livelihoods*. FAO. Kampala, Uganda.
Food and Agricultural Organisation (2020): *Kenya at a Glance*. At: www.fao.org/kenya/fao-in-kenya/kenya-at-a-glance/en/ [retrieved last 8 February 2020].
Food and Agricultural Organization Stat (2017): *Emerging Practices From Agricultural Water Management in Africa and the Near East. Water Harvesting*. At: www.fao.org/fileadmin/user_upload/agwa/docs/bari2017/Session3/Harvesting_Introduction_I.pdf [retrieved last 22 February 2020].
Government of Kenya (2010): *The Constitution of Kenya*. Republic of Kenya. Nairobi.
Government of Kenya (2018): *Kenya Development Response to Displacement Impacts Project (KDRDIP) Additional Financing. Environmental and Social Management Framework*. Republic of Kenya. Nairobi.
Government of Uganda (1995): *Constitution of the Republic of Uganda, 1995*. Republic of Kenya. Kampala.
Gumba, D. (2018, 2 August): Interview by Renner, J. Nairobi, Kenya.
Harper, D., Boar, R. R., Everard, M. and Hickley, P. (2002): *Developments in Hydrobiology: Lake Naivasha, Kenya*. Kluwer Academic Publishers. Dordrecht.
Herrero, M. T., Ringler, C., van de Steeg, J., Thornton, P., Zhu, T., Bryan, E., Omolo, A., Koo, J. and Notenbaert, A. (2010): *Climate Variability and Climate Change and Their Impacts on Kenya's Agricultural Sector*. International Livestock Research Institute. Addis Ababa.
International Panel on Climate Change (2014): *Climate Change 2014. Climate Change Impacts, Adaptation and Vulnerability*. Cambridge University Press. Cambridge.
International Panel on Climate Change (2019): *Climate Change and Land. An IPCC Special Report on Climate Change, Desertification, Land Degradation, Sustainable Land Management, Food Security, and Greenhouse Gas Fluxes in Terrestrial Ecosystems*. United Nations. Geneva.
International Panel on Climate Change AR 5 (2014): *Climate Change 2014. Synthesis Report*. United Nations. Geneva.
International Research Institute for Climate Change and Society (2020): *IRI Climate and Society Map Room*. At: http://iridl.ldeo.columbia.edu/maproom/ [retrieved last 9 February 2020].
Kaka, I. (2019, 22 August): Interview by Renner, J. Nairobi, Kenya.
Kamese, G. (2018, 17 August): Interview by Renner, J. Kampala, Uganda.
Kapila, S. (2018, 10 August): Interview by Renner, J. Lake Naivasha, Kenya.
Kelly, P. M. and Adger, W. N. (2000): Theory and practice in assessing vulnerability to climate change and facilitating adaptation. *Climate Change* 47 (4), 325–352.
Kenya Vision (2020): *About Vision 2030*. At: https://vision2030.go.ke/about-vision-2030/ [retrieved last 5 February 2020].
Kiboye, M. (2018, 6 August): Interview by Renner, J. Nairobi, Kenya.
Kimbowa, R. (2018, 20 August): Interview by Renner, J. Kampala, Uganda.
Kimeta, E. (2018, 10 August): Interview by Renner, J. Naivasha, Kenya.
Kimeta, E. (2019, 19 August): Interview by Renner, J. Naivasha, Kenya.
Kitamirike, J. (2019, 23 August): Interview by Renner, J. Kampala, Uganda.
Kitzio, G. (2019, 22 July): Interview by Renner, J. Kikandwa, Uganda.
Konbusingye, L. (2019, 5 August): Interview by Renner, J. Kampala, Uganda.

Leipold, B. and Morgante, F. (2012): The impact of the flower industry on Kenya's sustainable development. *International Public Policy Review* 7 (2), 1–31.

Local Development Partners' Group in Uganda (2020): *Vision 2040 and National Development Plans*. At: www.ldpg.or.ug/vision-2040-and-national-development-plans/ [retrieved last 6 February 2020].

Mbaisi, C. (2018, 15 August): Interview by Renner, J. Nairobi, Kenya.

Mekonnen, M. M., Hoekstra, A. Y. and Becht, R. (2012): Mitigating the water footprint of export flowers from Lake Naivasha Basin, Kenya. *Water Resource Management* 26, 3725–3742.

Methmann, C. and Oels, A. (2014): Vulnerability. In: Death, C. (Ed.): *Critical Environmental Politics*. Routledge. London.

Ministry of Finance (2015): *Medium Term Plan (MTP)*. At: http://e-promis.treasury.go.ke/portal/development-strategy/medium-term-plan-mtp/ [retrieved last 5 February 2020].

Ministry of Finance, Planning and Economic Development (2020): *Bahati Presides over NDP III Ministerial Consultations*. At: www.finance.go.ug/press/bahati-presides-over-ndp-iii-ministerial-consultations [retrieved last 5 February 2020].

Muhindo, J. (31 July 2019): Interview by J. Renner, Kampala, Uganda.

Muigai, P. (2019, 21 August): Interview by Renner, J. Naivasha, Kenya.

Mukwaya, P. I., Sengendo, H. and Lwasa, S. (2012): Enhancing security and resilience of low-income communities to climate change in growing cities: An assessment of flood management and planning regimes in Kampala City, Uganda. In: Scheffran, J., Brzoska, M., Brauch, H. G., Link, P. M. and Schilling, J. (Eds.): *Climate Change, Human Security and Violent Conflict*. Springer. Berlin, Heidelberg.

National Geographic (2013): *10 Myths About Geothermal Heating and Cooling*. At: www.nationalgeographic.com/environment/great-energy-challenge/2013/10-myths-about-geothermal-heating-and-cooling/ [retrieved last 10 February 2020].

National Planning Authority (2020): *Uganda Vision 2040*. At: www.npa.go.ug/uganda-vision-2040/ [retrieved last 6 February 2020].

Nelson, D. R., Adger, W. N. and Brown, K. (2007): Adaptation to environmental change: Contributions of a resilience framework. *Annual Review of Environmental Resources* 32, 395–419.

Njiru, B. N. (2012): Climate change, resource competition, and conflict amongst pastoral communities in Kenya. In: Scheffran, J., Brzoska, M., Brauch, H. G., Link, P. M. and Schilling, J. (Eds.): *Climate Change, Human Security and Violent Conflict: Challenges for Societal Stability*. Springer. Berlin, Heidelberg.

Nyangena, W. and te Velde, D. W. (2012): *Managing the Water-Energy-Land (Wel) Nexus for Inclusive and Sustainable Growth. A Case Study of Flower Production Around Lake Naivasha*. European Report on Development. Brussels.

Ongech, D. (2019, 27 August): Interview by Renner, J. Lodwar, Kenya.

Opiyo, F. (2018, 7 August): Interview by Renner, J. Nairobi, Kenya.

Opiyo, F., Nyangito, M., Wasonga, O. V. and Omondi, P. (2014): Trend Analysis of rainfall and temperature variability in arid environment of Turkana, Kenya. *Environmental Research Journal* 8 (2), 30–43.

Otieno, A. (2019, 14 August): Interview by Renner, J. Nairobi, Kenya.

Otieno, B. P. A. (2019): *Influence of Poverty on Violent Extremism in Kenya: A Case Study of Mombasa County Between 2007 to 2017*. United States International University-Africa. Nairobi.

Peter, K. T., Waswa, M. N. and Oloo, A. O. (2018): The unemployment puzzle an insurgence to insecurity in Nairobi County, Kenya. *Journal of Security Studies and Global Politics* 3 (2), 131–137.

Peter, S. (2018, 9 August): Interview by Renner, J. Naivasha, Kenya.

Peter, S. (2019, 19 August): Interview by Renner, J. Naivasha, Kenya.
Raleigh, C., Jordan, L. and Salehyan, I. (2008): *Assessing the Impact of Climate Change on Migration and Conflict. Working Paper for the World Bank, Social Dimensions of Climate Change.* At: http://siteresources.worldbank.org/EXTSOCIALDEVELOPMENT/Resources/SDCCWorkingPaper_MigrationandConflict.pdf [retrieved last 10 May 2019].
Ran, A. M. (2018, 14 August): Interview by Renner, J. Nairobi, Kenya.
Republic of Uganda (2010): *National Development Plan 2010/11–2014/15.* Republic of Uganda. Kampala.
Republic of Uganda (2015): *Second National Development Plan (NDP II) 2015/16–2019/20.* Republic of Uganda. Kampala.
Scheffran, J., Brzoska, M., Brauch, H. G., Link, P. M. and Schilling, J. (2012): *Climate Change, Human Security and Violent Conflict.* Springer. Berlin, Heidelberg.
Scheffran, J., Link, M. and Schilling, J. (2019): Climate and conflict in Africa. In: Claussen, M. (Ed.): *Oxford Research Encyclopedia. Climate Science.* Oxford University Press. Oxford.
Schilling, J., Akuno, M., Scheffran, J. and Weinzierl, T. (2014): On raids and relations: Climate change, pastoral conflict and adaptation in northwestern Kenya. In: Bronkhorst, S. and Bob, U. (Eds.): *Climate Change and Conflict: Where to for Conflict Sensitive Climate Adaptation in Africa?* Berliner Wissenschaftsverlag. Berlin.
Schilling, J., Locham, R., Weinzierl, T., Vivekananda, J. and Scheffran, J. (2015): The nexus of oil, conflict, and climate change vulnerability of pastoral communities in northwest Kenya. *Earth System Dynamics* 6 (2), 703–717.
Schleussner, C. F., Donges, J. F., Donner, R. V. and Schellnhuber, H. J. (2016): Armed-conflict risks enhanced by climate-related disasters in ethnically fractionized countries. *Proceedings of the National Academy of Science of the United States* 113 (33), 9216–9221.
Schmitz, C., Lotze-Campen, H., Gerten, D., Dietrich, J. P., Bodirsky, B., Biewald, A. and Popp, A. (2013): Blue water scarcity and the economic impacts of future agricultural trade and demand. *Water Resource Research* 49 (6), 3601–3617.
Schuyt, K. D. (2005): Economic consequences of wetland degradation for local populations in Africa. *Ecological Economics* 53 (2), 177–190.
Shem Otoi, S. (2016): Modelling economic determinants of youth employment in Kenya. *Journal of Emerging Trend in Economics and Management Science* 7 (1), 31–38.
Standardized Precipitation Evapotranspiration Index (2020): *Global Drought Monitor.* At: https://spei.csic.es/map/maps.html#months=1#month=8#year=2018 [retrieved last 20 February 2020].
State House of Uganda (2016): *Strategic Guidelines and Directives for the Term: 2016 2021.* At: www.statehouse.go.ug/media/speeches/2016/06/23/strategic-guidelines-and-directives-term-2016-2021 [retrieved last 6 February 2020].
Thomas, R. J. (2008): Opportunities to reduce the vulnerability of dryland farmers in central and West Asia and North Africa to climate change. *Agriculture, Ecosystem & Environment* 126 (1–2), 36–45.
Tindimungaya, C. (2018, 23. August): Interview by Renner, J. Kampala, Uganda.
Transparency International (2018): *Corruption Perception Index 2018.* At: www.transparency.org/cpi2018 [retrieved last 14 February 2020].
Twesigye, B. (2019, 06 August): Interview by Renner, J. Kampala, Uganda.
UKAID, PACT and Alliance for Responsible Mining (2018): *Economic Contributions of Artisanal and Small-Scale Mining in Uganda: Gold and Clay.* Westcombe. Envigado.
Uluocha, N. O. and Okeke, I. C. (2004): Implications of wetland degradations for water resource management: Lessons from Nigeria. *GeoJournal* 61, 151–154.

United Nations Department for Economic and Social Affairs (2020): *Population Dynamics. World Population Prospects 2019.* At: https://population.un.org/wpp/DataQuery/ [retrieved last 13 February 2020].

United Nations Development Programme (2019): *Global Knowledge Index.* At: www.undp.org/content/undp/en/home/search.html?q=Global+Knowledge+Index [retrieved last 8 February 2020].

United Nations Development Programme (2021a): *Human Development Index.* At: http://hdr.undp.org/en/countries [retrieved last 8 February 2021].

United Nations Development Programme (2021b): *Human Development Reports. Education Index.* At: http://hdr.undp.org/en/content/education-index [retrieved last 8 February 2021].

Vasquez, P. I. (2013): Kenya at a crossroads: Hopes and fears concerning the development of oil and gas reserves. *International Development Policy* 4 (3), 1–23.

Vivekananda, J. (2015): *Peace Audit Kenya.* International Alert. London.

Wamuyu, K. J. (2016): Challenges facing devolution in Kenya. In: *DSpace Repositorium.* Masai Mara University. Narok.

World Bank Group (2018): Poverty and shared prosperity 2018. In: *Piecing together the Poverty Puzzle.* World Bank Group. Washington.

World Bank Group (2019): *The World Bank in Kenya. Overview.* At: www.worldbank.org/en/country/kenya/overview [retrieved last 5 February 2020].

World Bank Group (2020a): *Agriculture, Forestry, and Fishing, Value Added (% of GDP) – Uganda.* At: https://data.worldbank.org/indicator/NV.AGR.TOTL.ZS?locations=UG [retrieved last 8 February 2020].

World Bank Group (2020b): *Climate Change Knowledge Portal.* At: https://climateknowledgeportal.worldbank.org/download-data [retrieved last 13 February 2020].

World Bank Group (2020c): *Climate Change Knowledge Portal. Uganda. Climate Data. Historical.* At: https://climateknowledgeportal.worldbank.org/country/uganda/climate-data-historical [retrieved last 9 February 2020].

World Bank Group (2020d): *Total Investment (% of GDP).* At: https://tcdata360.worldbank.org/indicators/inv.all.pct?country=KEN&indicator=345&countries=BRA&viz=line_chart&years=1980,2024 [retrieved last 5 February 2020].

World Bank Group (2020e): *World Bank Development Indicators.* At: https://databank.worldbank.org/reports.aspx?source=2&series=NY.GDP.PCAP.KD.ZG [retrieved last 5 February 2020].

World Bank Group. (2020f): *Manufacturing, Value Added (% of GDP)-Kenya.* At: https://data.worldbank.org/indicator/NV.IND.MANF.ZS?locations=KE [retrieved last 5 February 2020].

World Bank Group (2020g): *GDP Growth (Annual %)-Uganda.* At: https://data.worldbank.org/indicator/NY.GDP.MKTP.KD.ZG?locations=UG [retrieved last 6 February 2020].

World Bank Group (2020h): *Mortality Rate, under-5 (Per 1,000 Live Births) – Kenya, Uganda.* At: https://data.worldbank.org/indicator/SH.DYN.MORT?locations=KE-UG [retrieved last 13 February 2020].

World Bank Group (2021): *GDP Per Capita (Current US$)-Uganda.* At: https://data.worldbank.org/indicator/NY.GDP.PCAP.CD?locations=UG [retrieved last 6 February 2021].

World Wildlife Fund Employee (2018, 10 August): Interview by Renner, J. Naivasha, Kenya.

Yakub, Y. (2019, 02 September): Interview by Renner, J. Nairobi, Kenya.

5

WATER ACCESS AND (VIOLENT) CONFLICT

How Economic Development and (De)centralization Result in Water Insecurity

5.1 Conflict Dynamics Between and Across Local-Level Actors at Lake Naivasha and Lake Wamala

The conflict analysis is geared to the four-stage process as proposed by Ropers (2008). Firstly, it needs to be defined whether a conflict exists. Secondly, a detailed conflict analysis is conducted. This step includes the identification of the conflict parties, the conflict issues of the identified actors with respect to their positions, interests, values, needs, perceptions, powers and feelings (Galtung 1996; Ropers 2008; Ramsbotham et al. 2011). More so, the history and duration of the conflict, and structural and contextual features are described, which influence the conflict and determine its dynamics respectively (Poolman et al. 2009; Ramsbotham et al. 2011). The most frequent differentiation takes place between structural and proximate causes of a conflict. Structural causes are long-term or strategic causes which are built on norms, structures and policies within a political and social system. Proximate causes are happening more recently and can change quickly.

The data analysis (see Figure 4.4) illustrated that the risk over the outbreak of conflicts are the highest within the local-level stakeholder group. The figure, however, does neither present the breakdown of the identified stakeholders within the group of local-level actors nor provides information on how the conflict dynamics differ between Lake Naivasha and Lake Wamala. Respectively, a mixed-method case study analysis approach allowed to identify conflict risks between the local-level actors.

The data analysis method, firstly, provided in-depth information on the frequency distribution within the local resource users' stakeholder group. Secondly, the distribution of how the conflict risk differs between the two case studies was captured. The results of the conflict risk distribution between the local-level actors are set out in Figure 5.1. From the figure it can be seen that the conflict risk

DOI: 10.4324/9781003271291-5

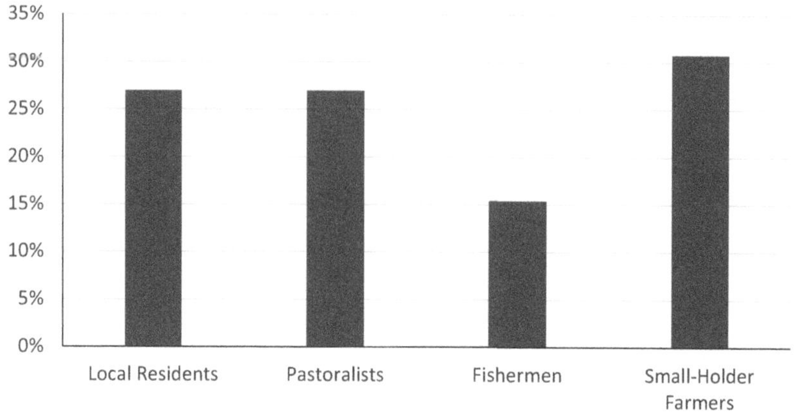

FIGURE 5.1 Conflict Risk Between Local-Level Actors

(*Source:* The Author 2020)

between the small-holder farmers is the greatest (31 per cent), followed by the local population (27 per cent) and the pastoralists (27 per cent). The conflict risk among the fishermen is the lowest (15 per cent). What stands out is the relatively high conflict risk between the pastoralist communities, even though the previous analysis demonstrated that pastoralists are hardly found around Lake Wamala. From Figure 5.2, it is apparent that the high number for pastoralists is due to the appearance of pastoralists at Lake Naviasha (34 per cent) compared to 7 per cent for Lake Wamala.

The great values for Lake Naivasha's local population derive from the increasing population numbers in the lake basin and especially in the informal settlements.

As such, more people push onto the four remaining landing sites or the few water access points in the informal settlement and, therefore, the demand is increasing, while at the same time access points are decreasing. Contrarily, the reported conflict risk between the local residents is almost non-existent at Lake Wamala. The reason for this has to do with the small amount of people living at the lake directly. More so, in the wider surroundings of Lake Wamala, the communities are often far away from each other so people hardly get together. Comparing the conflict risk between fishermen and small-holder farmers, the chance for a conflict to erupt is higher in both cases for Lake Wamala. There are two likely causes for the difference between the conflict risk. The offer of employment to unskilled workers in the flower farms and hotels is a reason why former small-holder farmers turn to other income opportunities. The merger of fishermen in fishing associations slows down the conflict risk between the fishermen at Lake Naivasha. The two latter cases can be ruled out for small-holder farmers and fishermen at Lake Wamala.

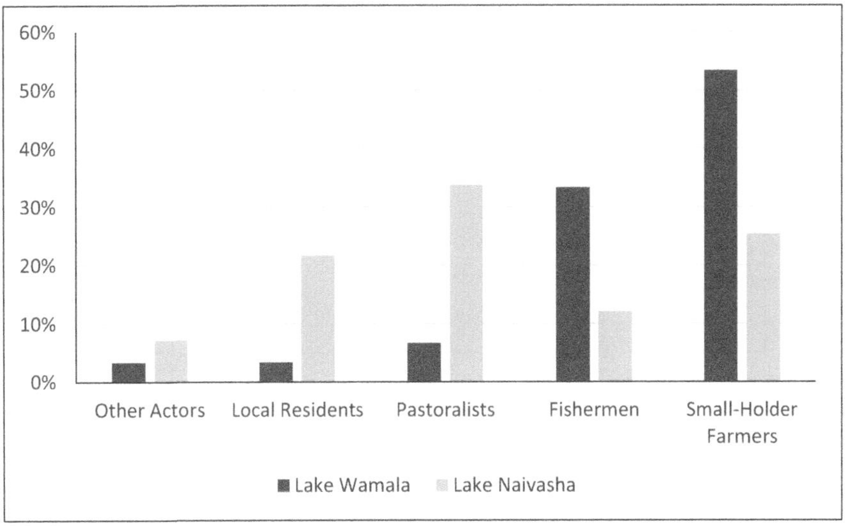

FIGURE 5.2 Comparison of Conflict Risk Between Local-Level Actors at Lake Naivasha and Lake Wamala

(Source: The Author 2020)

The data and the recent discussion of the social implications on the conflict risk provide further support for conflictual tensions, not only within the group of small-holder farmers or local residents but also in between these groups. Figure 5.3 shows that the highest values for intergroup conflicts are between pastoralists and small-holder farmers (32 per cent) and among fishermen and small-holder farmers (40 per cent). The lowest values are reported for pastoralists and fishermen (8 per cent). From the context analysis, the situation seems similar for both lake sites. Nevertheless, the data analysis confirms differences regarding the conflict risks between the various local resource user groups. Therefore, the following section discusses the conflict situation between small-holder farmers and pastoralists for Lake Naivasha and the conflict situation associated amidst small-holder farmers and fishermen at Lake Wamala.

5.2 Small-Holder Farmer-Pastoralist Conflict at Lake Naivasha

5.2.1 Conflict Actors (Positions, Interests, Needs)

As described on the previous pages, conflicts already exist between local resource users. This part of the section is guided by the overarching question – who are the actors that influence conflict dynamics? Drawing on Fisher et al. (2000), primary actors are defined as those who have direct impact on the conflict. Secondary actors

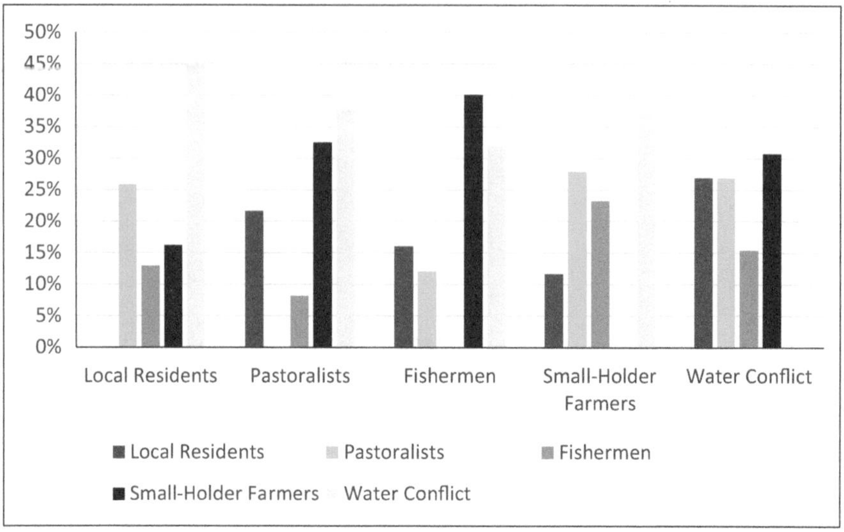

FIGURE 5.3 Risk Over the Outbreak of Water-Related Conflicts Across the Local-Level Actors

(Source: The Author 2020)

are characterized as those who have indirect impact on the conflict. Lastly, tertiary actors are either those who the conflict has an impact on, as those who engage in peace activities or those who act as spoilers[1] (Fisher et al. 2000). The primary conflictual actors are small-holder farmers and pastoralists. The conflict situation is influenced by both national and regional political stakeholder's decision-making processes, as well as national and international economic actors. Further, fishermen and local residents around Lake Naivasha basin can be classified as tertiary actors.

Both small-holder farmers and pastoralists have competing interests and must make trade-offs. Small-holder farmers take the position that they want to uphold the land authority over the few remaining farming areas without sharing those with other local resource users (see Table 5.1). Furthermore, they insist on the openness of water access routes to the lake's shorelines. They advocate the position that either public farmland remains free for them to use or they demand for the legal use of land within the catchment areas. Contrarily, pastoralists claim authority over public and free land as grazing areas. Additionally, they insist on unprotected water access routes to water their cattle. They argue that they should be allowed free movements, as they are only around for a couple of days before moving further. Regarding the question of employment, pastoralists ask for financial benefits from the improved infrastructural developments as they lost land through the set-up of economic sites and road constructions. Likewise, they claim jobs in the newly set-up economic sites. The latter holds true especially for the Maasai who lost land around Lake Naivasha or the Samburu in Laikipia county.[2] In a similar vein,

TABLE 5.1 Positions, Interests and Needs of Primary Conflict Actors at Lake Naivasha

Small-Holder Farmers		*Pastoralists*
Uphold land authority over remaining farming areas, claim over free water access and free farming land	**POSITIONS** **What one *says* they want**	Financial benefits from improved infrastructural developments, claim over grazing authority and water access
Land redistribution, improved water access and infrastructure, increased access to health and education services, employment	**INTERESTS** **What one *really* wants**	Migration routes, water and land access, employment, financial compensations, education and health access
Access to land and water, wellbeing and security	**NEEDS** **What one *must* have**	Land and water for grazing, wellbeing and security

(*Source*: The Author 2020)

Schilling et al. (2015) noted that the Turkana or Pokot claimed financial compensations and employment from Tullow in Turkana County (Schilling et al. 2015).

The small-holder farmers' interests can be derived from their dependence on available land and water to pursue their agricultural activities. Consequently, the interest of having a fair and equal land distribution, as well as access to improved water points between all stakeholders is pursued. Thus, they are interested in a secure and stable relationship with the other local resource users. Apart from the resource security aspect, they strive for better access to health and education facilities. Access to better education is especially of severe importance as it depicts their potential for an opportunity to get better employment in one of the economic sites. Thus, this interest in education and stable resource access is of crucial importance, as it is directly linked to their need of security. The reintroduction of the traditional rights which regulated the seasonal use of grazing areas by pastoralists in this region and beyond is of significant interest for the pastoral communities. Beyond that, pastoralists have an interest in good relations with other local stakeholders, not least for the reason of accessing land and water. In terms of economic interests, pastoralists are keen to receive financial compensations for the lost land. In a similar vein to small-holder farmers, they are interested in improved education and health centre services. Education is seen as a need to be given a choice of whether they want to resettle to other regions or look for skilled employment opportunities. Education might also help them to better understand the existing laws and to negotiate for better conditions with the economic stakeholders. All mentioned interests are, however, directly linked to their need of security and further general wellbeing.

The positions and interests have to be understood against a wider background to exert power. There is a strong perception among both the pastoral communities and the small-holder farmers that they have been 'forgotten' (Balama 2019;

Community Member Karagita 2019) by both the central government and the county representatives. Some small-holder farmers residing in Karagita argued that community-based and non-governmental organizations (spoilers), who formerly expressed critical voices towards the economic investments undertaken, have been bribed with water access points on their private lands or have been granted senior positions and, therefore, remain mostly silent nowadays. Furthermore, the small-holder farmers and pastoralists' low level to exert power is also driven by the companies and governmental representatives' high capabilities to exert power over political institutions. Whereas these secondary actors have both financial means, relations to and support of the national government, small-holder farmers and pastoralists argued that they have been 'chased away' (Kapila 2018; Balama 2019) from their land by the economic actors who fenced off most public farming and grazing areas (Kapila 2018). As a result, small-holder farmers and pastoralists can only observe the flower farms and other economic actors coming in with heavy and expansive machinery as well as with water purification plants, building roads, drilling water on their private premises or constructing water infrastructures at the economic sites or private land that both pastoralists and the farmers are unable to access. The mentioned perceptions, the low level of resilience and the high economic developments undertaken in Lake Naivasha basin create a hotbed for extremely high inter-community hostilities. The latter is even accelerated as pastoralists move around armed. Majority of the respondents commented that pastoralists purchase guns from the conflicting bordering regions of Kenya to South Sudan, Ethiopia or Somalia. Accordingly, it was reported that pastoralists are able to use weapons to protect their cattle. Small-holder farmers, however, are left with their daily farming gadgets or provisional fences to protect their agricultural land. In contrast to small-holder farmers, pastoralists seem to have more power to assert their positions and interests towards small-holder farmers.

As small-holder farmers have only limited economic and social influence on resource access or land and water distribution, they do not seem to have noteworthy capacities to affect the conflictual setting in their favour. Their low capacity can be related to their permanent residence around the lake and are, therefore, easily traced in case of demonstrations or conflictual behaviour towards the secondary actors. More so, as some small-holder farmers also work in the flower farms to earn an additional income, they reported to be relieved from their obligation if they demonstrate against working-conditions or the closing of access routes. Thus, this explanation may also be attributed to the small-holder farmers low capacity to affect the conflict context. The only alternative small-holder farmers have to alter the conflict context is by fencing off their farmland or by guarding against trespassing of pastoralists and their cattle.

Similarly, pastoralists' capacity to influence the conflict setting is considered to be low. With regard to their relationship to secondary actors, pastoralists argued that neither the national government and country representatives nor the economic stakeholders offered them any employment or financial compensations, despite demonstrations and expressed claims for financial compensation. Thus, the pastoralists' expectation towards almost all secondary actors is characterized by frustration. The capacity of pastoralists to change the conflictual setting towards small-holder farmers can be described to be higher than their opponent's. A possible explanation for this is the short-term stay of the pastoralists in the area and, therefore, they are more likely to be involved in the conflictual behaviour. Moreover, pastoralists are used to conflictual situations, as livestock raiding and the usage of semi and fully automatic small arms became part of pastoral life and still characterizes intra-pastoral conflicts nowadays (see, e.g. Adem et al. 2012; Ember et al. 2012; Schilling et al. 2012). As a result, pastoralists are more likely to affect the small-holder farmers vis-a-vis the pastoral conflict setting in their favour through the use of arms.

When asked about possible incentives for peace, both groups stated that they do not foresee any improvement of the situation in the short term. Historic grievances (see sub-section 5.2.2), ethnic affiliations and also prestige are possible incentives for conflict for the Maasai and Samburu. The mentioned incentives for conflicts are said to be similar for small-holder farmers. Added to that, small-holder farmers also argue that economic motives (in form of crops) are reasons for the continuation of conflictual behaviour. In addition to secondary actors who influence the conflict setting through economic and political kinship, the other local-level actors can be considered tertiary actors who influence the conflict dynamics (see sub-section 5.2.3). Because of the political ties and the funding that community-based and local non-governmental organizations receive from national actors and also partly from international organizations, they might be considered potential spoilers who could divide the local-level actors even further.

Hostile perceptions among the small-holder farmers and the pastoralists can be considered an implication of the baseline analysis of the conflict actors and their characteristics respectively. As it will be discussed in further detail later, the perceptions between the primary conflict actors towards the tertiary actors can also be described as hostile, while the perceptions of the primary actors against secondary actors range from hostile nowadays to cautiously hopeful in the mid-term.

5.2.2 Conflict Profile and Conflict Causes (History and Structural and Proximate Causes)

The overarching questions discussing the conflict profile are, firstly, what causes and caused the conflict and, secondly, what is the context that shapes the conflict? As explained in Chapter 3, Naivasha basin was historically a Maasai grazing region and its main industry was foremost farming and fishing (Kundu et al. 2010). At the present time, there are a number of economic activities, resulting in a complex situation 'as there are different users from different locations in different economic sectors' (Vivekananda 2015: 45). An employee from Lake Naivasha Water Resource Users Association (LNRUA) stated accordingly that the area is 'a very cosmopolitan area with a lot of different tribes[3] with different needs [that] need to be balanced'. With the population increase, the grievances, especially between young men from opposing tribes and from different economic backgrounds (i.e. small-holder farmers, pastoralists or fishermen), have been the strongest. Taking a closer look at the existing conflict data for Lake Naivasha basin, the conflict profile has mostly been local, sporadic and of low intensity without a direct involvement of government and government security forces[4] (see Figure 5.4).

Alongside the indicated higher levels of conflict between local-level actors, the analysis of the population development affirms the settlement of members of the Kikuyu tribe around Lake Naivasha after independence. While they practised rain-fed farming mostly, the incoming fishermen primarily belonged to the Luo tribe. As a result of environmental destructions, over-abstractions in the upper catchment areas and climate change, the amount of cattle increased in the basin, as both Maasai and Samburu came closer to the lake shores. As indicated in the previous section, inter-tribal grievances have been mentioned as a reason for the continuation of conflictive behaviour among the pastoralists, but also between small-holder farmers and pastoralists. Previous studies analysed the structures behind the animosities (see, e.g. Shettima and Tar 2008; Goldsmith 2013; Schilling et al. 2014). Ahead of the arrival of economic stakeholders, Kenya's government has conferred preferential land rights to members (often small-holder farmers) of their own ethnic group, thereby marginalizing any other 'foreign' group. Changes in governmental power

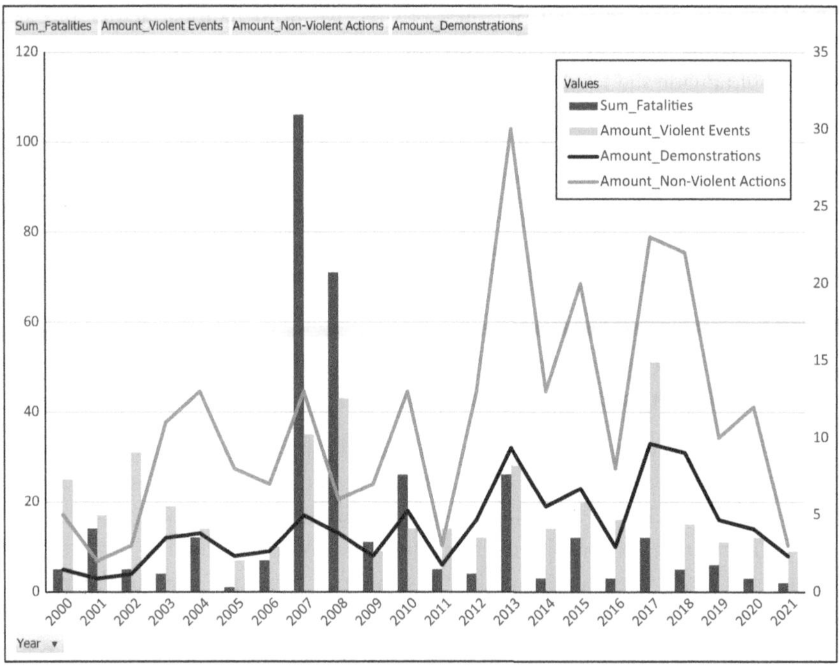

FIGURE 5.4 Timeline of Conflict Events at Lake Naivasha Since 2000

(*Source:* The Author 2021 based on ACLED 2021 and SCAD 2021)

coincided with the empowerment of the 'other' group and, thereby, the social tensions between the groups over land rights or resource accessibility increased. Respectively, reported violent and non-violent events (Figure 5.4) frequently revolved around issues concerning land use, in addition to disputes over water and livestock raiding (Ibid. 2014; ACLED 2021). Livestock raiding, especially, is perceived as a conflict issue that weighs heavily between the different livestock owner groups. Inter-pastoral conflicts have been there for a long time and are still accelerating existing conflict as it is perceived as a good means to solve past issues (Schilling et al. 2014).

Furthermore, the residents from Karagita were unanimous in the view that there are higher insecurity levels in the informal settlements. The residents articulated their concern that the arrival of people from other parts of the country increases the levels of insecurity in the informal settlement. Thus, they linked the high crime rates to the new incomers and their lack of employment. A few respondents identified youth employment as the biggest challenge the communities face and, therefore, argued it to be the biggest conflict driver in the basin. When asked about past conflict drivers, interviewees mentioned continued acts of violence by youth gangs. This statement is supported by the data retrieved from both ACLED and SCAD. However, caution must be applied as the findings have only been communicated by a small sample size. Violent events in the form of rioting ensue after sexual exploitations in the flower farms. Furthermore, after a boy drowned in an irrigation place, there have been some violent clashes between the Maasai and a flower farm. The Maasai held the flower farm responsible for the death of the boy as the channel was constructed by the flower farm. Non-violent events included pastoralists protesting against the drilling of geothermal vales and resulting fears of environmental and health damages, inter-tribal clashes after the attempts of cattle stealing or demonstrations against the government for its inactivity to follow up on the death of a local resident who seemed to be shot while trespassing a private land to access the lake (ACLED 2021; SCAD 2021).

In summary, the historical aspects indicate that both structural (root causes) and proximate motivations (immediate causes) caused the conflict profile. Structural causes, including political structures and affiliations, as well as economic growth processes, shaped the conflict dynamics between the primary, secondary and tertiary actors around Lake Naivasha. Social aspects, such as ethnic affiliations, characterized the conflict dynamics between the primary conflict actors, especially after Kenya gained independence. This, for example, resulted in unequal land distribution and fuelled inter-ethnic grievances. After the economic potential of the area (proximate cause) was discovered, institutional reform processes and economic growth agendas shifted the conflict profile. Land and water resources have been used by the economic actors increasingly. Resulting thereof, primary actors demonstrated against the secondary actors' resource exploitation because their economic undertakings reduced the local-level actors' access to water and land resources. However, since the turn of the millennium, violent and non-violent events decreased because local-level actors feared job or water supply constraints (structural causes). Whereas the violent and non-violent behaviour of the primary actors towards the secondary actors slowed down, the conflictual situation between the different local-level actors intensified. Since the beginning of the 2000s, people migrated to Lake Naivasha to either graze their cattle, set up agricultural activities or in search of labour, social amenities and natural resources. More so, the overspill of violent conflict and arms proliferation from the neighbouring countries (Somalia or South Sudan) reinforced internal migration because of violent attacks by terrorist groups in the border region (e.g. Al Shabab) (proximate cause).

To sum up, political power contestations shaped inter-ethnic rivalries after independence. This still serves partly as a conflict trigger between small-holder farmers and pastoralists nowadays.[5] Moreover, economic developments accelerated inequalities and resulted in social and economic marginalization of especially the primary and tertiary conflict actors because they are still dependent on natural resources to sustain a living. The prioritization of economic actors and the lack of county and local authority over decision-making processes accelerate political exclusion and decreased resource access for local-level actors. The causes of conflictual behaviour are deeply rooted and politically entrenched (see Figure 5.4). However, conflictual tensions were increasingly concentrated at the local level. Taken together, these results suggest that the conflict profile over water alienation is shaped without direct governmental and security force involvement (exception: post-election violence in 2008, 2009) and takes place between the primary conflict actors. Nevertheless, the resource demand of secondary and tertiary actors triggers the conflict dynamics between the primary conflict actors.

5.2.3 Conflict Dynamics

Having previously discussed the conflict actors, the profile and causes of the conflict, this section moves on to examine the current conflict dynamics and trends between small-holder farmers and pastoralists. The considered dynamics result from the interaction of the conflict history, its actors and causes and how they can be triggered by current ongoing events (Herbert 2017).

The current conflict dynamics and the anger of the small-holder farmers against the pastoralists and vice versa are not much driven by the externalities of the resource exploitations of the flower farms, hotel sites and the geothermal companies. The analysis of the conflict trends revealed that unmet expectations by the small-holder farmers and pastoralists are recent conflict drivers. As it can be seen from Figure 5.4, the quantity of violent and non-violent events over the course of the past 20 years was around two to four events per year. This result, however, is somewhat counterintuitive. The most striking observation to emerge from the two research phases and the comparison with the conflict data from ACLED or SCAD was the high number of conflictual tensions which took place (and are still taking place) without external perception.

During the first research visit, a resident[6] from Naivasha described the situation as 'creating a lot of social tensions' resulting in 'disputes which do not necessarily mean that people are having a real fight'. Without being asked about it, participants of a mixed gender focus group discussion mentioned that the flower farms and hotels are increasingly viewed as stressors and an enemy due to their pollution of the water or their unpredictability in offering labour or the handling of its employees (dismissal or sexual exploitations). During the second research phase, the community members and residents of Naivasha spoke of another stressor which worsens the general situation for the local-level actors: geothermal companies and the cargo-centre which is going to be constructed. The majority of the interviewees supported the opinion that the conflict dynamics between the local-level actors, especially between pastoralists and small-holder farmers and between the residents in the informal settlements, has built up generally.

Resulting thereof, their situation can be characterized as being caught between multiple enemies, that is, on one side, there are economic companies and political actors, while on the other side, there are local-level actors and external (Kenyan) actors migrating to the lake. Between the two research visits, the level of non-violent and violent acts between the small-holder farmers and the pastoralists had increased. When asked about the rise in inter-communal conflicts, the communities were unanimous in the view that demonstrations, strikes for better working conditions or protests against the closure of a public corridor leading to Lake Naivasha result in dismissals or change in water ratios in the installed water tanks. Hence, non-violent and violent acts against the companies are no longer considered promising activities to change the situation. This, in turn, however, accelerated the level of conflict among the local-level actors, especially between the primary conflict actors.

The increasingly violent behaviour between the primary actors undermines the need for conflict resolution and, therewith, reinforces the need for violent acts to enforce one's own right to water and land access. Horizontal economic and political inequalities, furthermore, increases the risk of conflict as a result of, firstly, the absence of effective formal regulations and a general lack of government involvement in the local-level actors' concerns. Secondly, economic growth, as set out in the Vision 2030, continues to be driven by investments in infrastructure, constructions, financial services and rapid urbanization, and consequently, does not meet the needs and level of education and knowledge of the primary conflict actors. Thirdly, the society in Kenya remains extremely stratified and political kinship and economic and political privileges fuel grievances between the local-level actors and community-based organizations or county government representatives as public access points are closed off or because of the economic benefits employees receive. A resident summarized this perception by stating 'it is a matter of the rich really fucking over the poor. The businesses are the rich. And [there is] a corrupt government and they all [including the county government and community-based organizations] work very well hand in hand'. He, furthermore, mentioned that inter-communal riots only happen because 'people are screwed over'. President Uhuru Kenyatta's declared war against corruption to make his legacy (BBC 2018) is seen by many only as a public activity to secure funding from international donors and to attract more international businesses.

The farmer-herder conflict and conflicts between the primary and tertiary actors are triggered by climate change. For instance, increased drought frequency and heavy rainfalls force nomadic herders to change their itineraries and compete for water and land with the small-holder farmers, especially at the four remaining landing sites. More so, cattle are also increasingly grazing along the road, causing road accidents. This, furthermore, affects the wildlife corridor between Hell's Gate National Park and the ranches along the lake. The gradual changes in weather conditions open up new opportunities for cultivation and incite agricultural encroachment onto pastoral land and the catchment areas. In the absence of effective regulations and conflict mitigation mechanisms, this might also lead to communal violence in the short to mid-term. Commitments to pastoralists and small-holder farmers by state and regional authorities have often remained unfulfilled. As a result, historically grown grievances, which caused pastoralist and farmer-herder conflicts in the past, could not have been addressed and continue to trigger the conflicts in recent times.

Additionally, spoilers have the power to change the intensity or direction of violence in the current conflict setting. In the conflict under consideration, community-based and non-governmental organizations can be considered spoilers. Interviewees reported that over the course of the past few years, community-based organizations decreased their active commitments to advocate for land and water rights for the primary and tertiary actors. During the field visits, employees from community-based organizations and NGOs explained that they try to provide water for the communities living nearby or to press for the implementation of existing laws. However, they highlighted that they do not have the power and influences in 'setting up good policies for [the] people . . . who are living around'. During the first research visit, meetings with NGO representatives took place at Lake Naivasha. During this meeting, the employees committed themselves to advocate for the enforcement of environmental laws and to put pressure on the local authorities to create better living conditions for local-level actors. The second research visit revealed in several ways that their effectiveness is limited to certain actor groups (e.g. advocating for fishermen rights given their contribution to the economy) or for their own interest only. As such, NGOs or employees of community-based organizations received permits or land titles from National Environmental Management Authority (NEMA) in case they work within the legislative framework provided by the national authorities, thereby promoting the economic development of the basin during public participation meetings. Furthermore, the general analysis pointed out that fishermen are wealthier than the other actor groups because they organize in fishing associations. Interviewees reported that the temporary 'wealth' of the fishermen can spark violence, either because of other local actors' greed for money or because the fishermen use the money to pay unmarried and married women for sexual favours. This especially fuels intercommunal and inter-ethnic grievances and accelerates existing social tensions.

A further factor contributing to the existing tensions might be the appearance of shadow actors. Shadow actors are not necessarily visible or impacting on the conflict situation directly, but their support for conflict actors can correspond to the division of conflict actors. The research findings point out that rebel groups and militias in the neighbouring countries (e.g. attacks by Al Shabab resulting in ongoing internal migration), international economic actors (especially Chinese investors) and the Governments of Uganda (involvement in the construction of the cargo centre) and Ethiopia (offering economic benefits attracting the flower farms at Lake Naivasha to move to Ethiopia) can be considered shadow actors.

A long history of partly non-violent and partly violent conflict between the farmers and herders, and also with other tertiary actors, has provided all relevant local-level stakeholders in the Lake Naivasha basin with the skills, knowledge and arms (pastoralists only) to execute violent attacks. Economic actors do not seem motivated to smoothen the situation or to adjust their operations by adapting corporate responsibility measures. Similarly, national and county representatives do not show interest in calming down the situation. Neither do they seem motivated to address underlying social security issues including access to health care or education facilities. Instead, their priority remains with the implementation of the economic development agenda, and up until now, the profits and revenues from the economic sites do not trickle down to the local level. Against this background, the conflict dynamics between the small-holder farmers and the pastoralists are significant. In addition, the conflict between the tertiary actors and the primary actors also has implications for the stability in the region as it fuels a further escalation of the already-existing violent tensions. The next section will move on to map the relationship of the conflict actors.

5.2.4 Conflict Mapping

The map represents the conflict geographically by placing the parties in relation to both the conflict issue and to the conflict actors. The map, moreover, helps not only to visualize the power alignments but also in mapping the needs and fears. The power relationships become evident through the relative size of the actors in the map (see Figure 5.5) and the lines between the actors symbolise the type of the relationship (e.g. alliance, or conflict over a particular issue). Mapping the conflict dynamics also allows outsiders to the conflict context to understand the overall conflict system and the relationship between the different conflict actors. However, a conflict map can show a particular moment in a specific situation only. Additionally, conflict mapping is dynamic and the map reflects a particular point in a changing situation and points towards an action. Furthermore, the map should not show objective aspects only, but it should also be useful to outline the issues between the parties that are in the conflict. For the aforementioned reasons, the layout of the map depends on the person who prepares it and, therefore, captures the perspective and point of view of the conflict researcher.

Figure 5.5 shows the mapping analysis of the pastoralist and herder conflict from the perspective of the field visits conducted in between 2018 and 2021 especially. It illustrates the conflict between any pastoralist group (Maasai or Samburu) and the small-holder farmers. Looking at the map, it is possible to identify the main parties

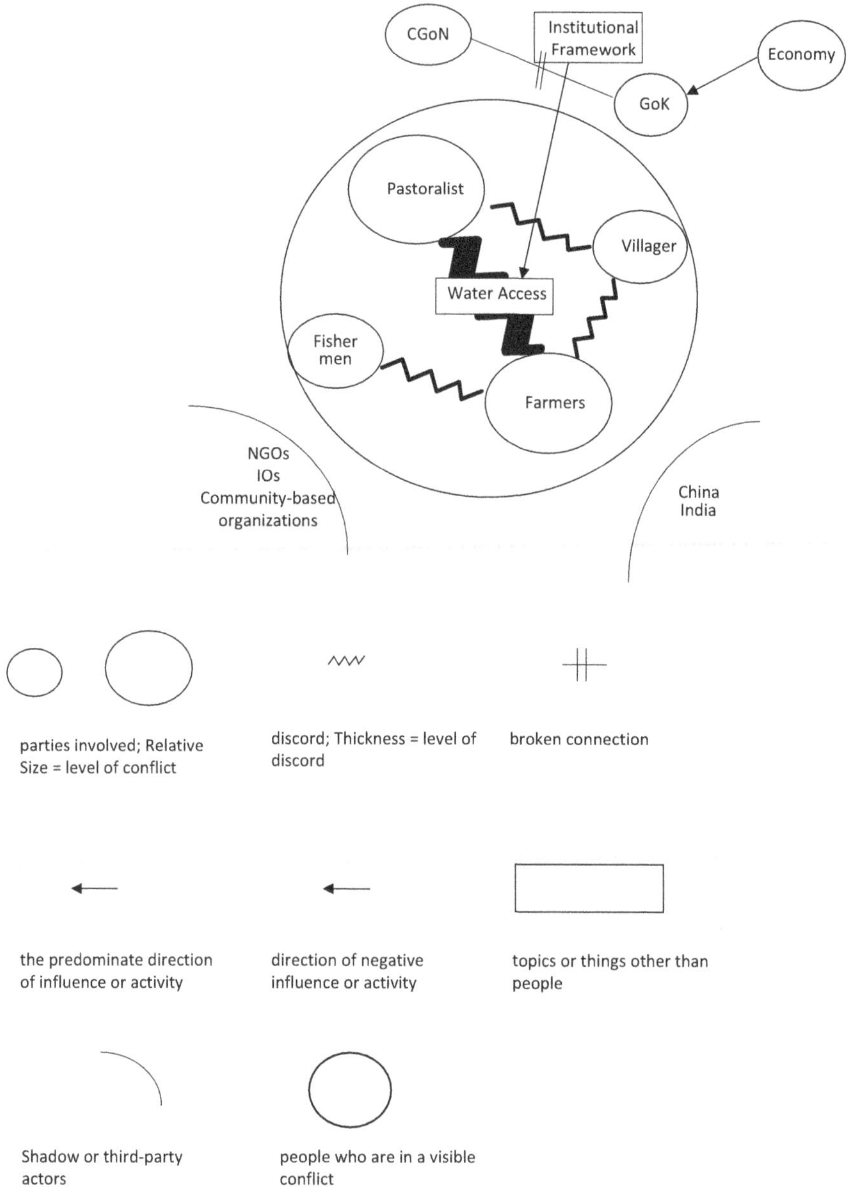

FIGURE 5.5 Conflict Map, Pastoralist-Farmer Conflict, Summer 2021

(*Source:* The Author 2020)

involved in the context, as well as the relationship between them. The relative size of the circle indicates the level of conflict between the parties. While the conflict relationship between the pastoralists and the farmers (represented by the bold zig-zag line) is the core issue, in the context and the basis for violent and non-violent responses to the conflict dynamics, other relationships are also important and are brought into focus when analysing the conflict situation. Thus, both fishermen and villagers negatively influence the conflict situation as both groups of actors are in conflict with the pastoralists (villagers) and the small-holder farmers (fishermen and villagers). The large dashed circle presents the conflict actors who are visibly at odds with each other. It was observed that the primary and tertiary actors are in conflict, whereas secondary actors (GoK and the County Government of Nakuru [CGoN])[7] influence the conflict context negatively. The broken connection between the GoK and the CGoN results from the ineffective implementation of the policy rights as laid out in the institutional framework. Economic actors (both national and international ones) influence the national actors to support their business activities in the basin, which in turn increases the broken connection between the GoK and the CGoN. Other external actors, such as NGOs, international organizations, community-based organizations or countries, do not seem to influence the situation directly. However, they should be included to show, firstly, that they are indirectly involved and, secondly, to draw conclusions about intervention opportunities. Lastly, the map further highlights that the topic associated with the broken connection and visible conflict dynamics are different. Whereas the broken connection between the secondary actors results from the institutional framework, the conflict dynamics between the primary conflict actors are associated with access, especially denied access over water.

5.3 Small-Holder Farmer and Fishermen Conflict at Lake Wamala

5.3.1 Conflict Actors (Position, Interests, Needs)

The research area is affected by inter-communal conflict between the primary conflict actors, small-holder farmers and fishermen. National actors influence the conflict situation through decision-making processes indirectly. Furthermore, national economic actors can be classified as secondary actors as they influence the primary conflict situation by setting up economic sites and environmental destructions. International organizations and NGOs shall be characterized as shadow actors, given their indirect role in decision-making processes and conflict interventions mostly through financial contributions.

Similar to Lake Naivasha, small-holder farmers and fishermen have competing interests over the land and water resources at Lake Wamala (see Table 5.2). Small-holder farmers want to uphold authority over the farming areas which are still accessible at Lake Wamala directly. Moreover, they claim access rights to the lake's shorelines through one of the three remaining fishing villages. Fishermen advocate

TABLE 5.2 Positions, Interests and Needs of Primary Conflict Actors at Lake Wamala

Small-Holder Farmers		*Fishermen*
Uphold land authority over remaining farming areas Claim over free water access and free farming land	**POSITIONS** **What one *says* they want**	Uphold right of abode in one of the three fishing villages Free access to the lake, stop farmers from polluting the lake
Compensation for loss of land, improved water access and infrastructure in new living areas, access to health and education services, employment	**INTERESTS** **What one *really* wants**	Access to the market to sell fish, financial compensation in case of relocations to wider surrounding of the lake, education facilities and improvement of the lake's access routes
Access to land and water, wellbeing and security	**NEEDS** **What one *must* have**	Water access for fishing, market access, security and wellbeing

(*Source*: The Author 2020)

the position that they are the rightful owners of the land which the three remaining fishing villages belong to. Additionally, they insist on free access to the lake's shorelines to be able to perform their fishing obligations. Furthermore, fishermen argue that small-holder farmers pollute the water as they use pesticides and chemicals to increase the crop productivity. Thus, they indicate that they want small-holder farmers to leave the land and to move into the wider surrounding of Lake Wamala.

The current interests of both primary actors can be derived from the progressively occurring relocations of both actor groups. As a result of the economic activities, both primary actor groups are relocated as the land is taken by the government or investors and sold to national and international businesses. Thus, small-holder farmers demand either financial or physical compensations for the loss of their land. Small-holders move to mostly uninhabited land. Consequently, they are interested in improved water access, infrastructure, as well as access to education and health facilities. The interest in education is of crucial importance as it is directly linked to their need of security and wellbeing. As it is expected that the effects of climate change will severely impact small-scale farming, small-holder farmers are interested in finding other employment opportunities. However, looking for skilled employment opportunities requires access to education facilities. Fishermen are interested in remaining in their fishing villages and are also keen to access the local market to sell their fish. Moreover, they are interested in improving lake access routes as trucks may be able to stop at the fishing villages, giving the fishermen the opportunity to sell to the traders directly. Fishermen are also interested in financial compensations once they have been relocated into the wider surroundings of Lake Wamala. Beyond that, fishermen also have an interest in improved access to education and health facilities allowing them to look for skilled employment opportunities. All mentioned interests are linked to their need of security and general wellbeing.

Neither small-holder farmers nor fishermen indicated that they expect jobs, revenues or any other benefits from the economic actors who set up their sites around the lakes. They argue that international companies come in with their own labour force and, therefore, are not in need of other employees. In addition, they do not count on the national or district government to look for manpower at Lake Wamala and its surroundings. There is a strong perception among them that they either have been 'forgotten' or remain unknown to the political stakeholders. Thus, their capabilities to exert power for their positions or interests vis-a-vis the economic and national stakeholders are close to zero. Hence, hostile perceptions among the small-holder farmers and the fishermen can be considered an implication for their lack of power, capabilities and knowledge to strive for a better implementation of their rights to access resources and basic amenities, including education, health and market facilities.

5.3.2 Conflict Profile and Conflict Causes

Figure 5.6 shows the conflict dynamics for Lake Wamala and Mityana. Similarly, the conflict profile is mostly local, sporadic and also of low intensity without direct involvement of the government and security forces. Between 2000 and 2021, small-holder farmers and fishermen expressed their level of insecurity by demonstrations and non-violent attacks. Interviewees reported on low death rates[8] over the period considered. A number of interviewees in mixed-gender focus group discussions and also during individual face-to-face interviews have indicated that demonstrations and violent events were used vis-a-vis the district and governmental officials to claim for the maintenance of the landing sites, land compensations or infrastructural developments. Since 2010, political and economic actors increased their interests and investments in the basin. The rise in demonstrations by the local communities against the loss of land and environmental destructions can be attributed to this. In addition to the protests against the governmental officials, there are increasing tensions between the primary conflict actors. Respondents argued that the conflict dynamics between the local resource users increased especially after political actors gained interest in the area and sold land to economic actors (proximate cause). The stagnated institutional reform process reduced the adaptive capacity of the farmers and fishermen and accelerated their competition for farmland and access to either the lake's water or pumping stations especially (structural causes).

In summary, increasing economic developments accelerated the poor conditions which the farmers and fishermen have experienced for several decades. However, prior to the beginning of the economic investments, neither small-holder farmers nor fishermen mentioned that they have been in conflict with each other. Thus, the causes for the responses to the conflict dynamics are, to a lesser extent, deeply

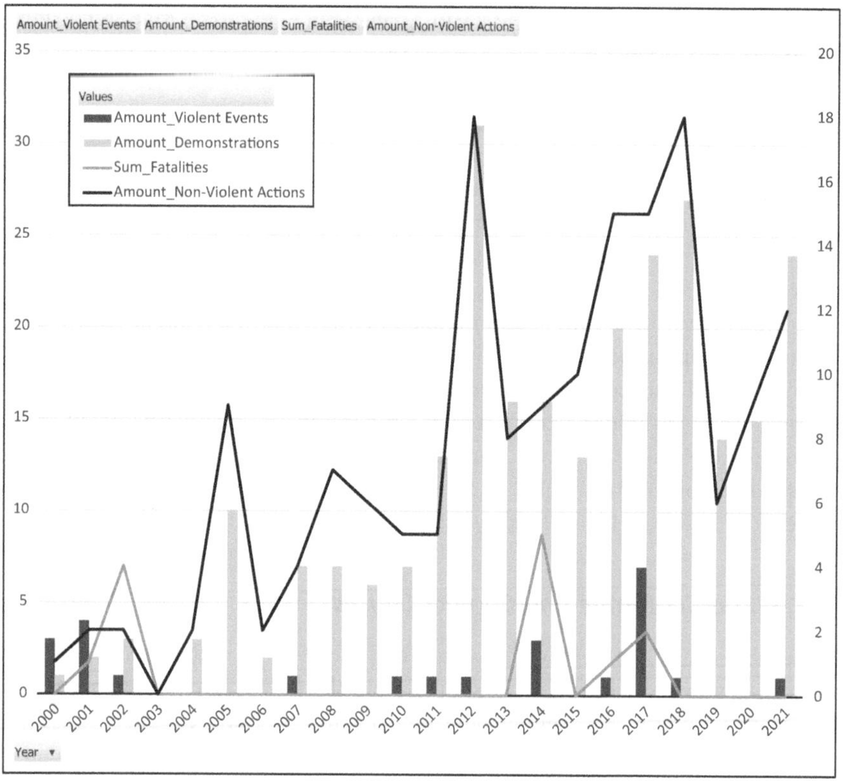

FIGURE 5.6 Timeline of Conflict Events at Lake Wamala Since 2000

(*Source:* The Author 2021 based on ACLED 2021 and SCAD 2021)

rooted or have been based on historically grown grievances[9] than they are the result of attempts to cope with the changing situation and to sustain a living given the new living environment.

5.3.3 Conflict Dynamics

As briefly touched upon, the current conflict dynamics result from the inability to cope with the currently existing environment. As indicated, farmers and fishermen felt that they were forgotten for a long time by the political elites. Respectively, their level of education and their knowledge about the institutional framework and the existing laws are low, or rather close to zero. Since the increasing investments and economic developments resulted in the relocations of farmers and fishermen, community-based organizations and a few environmental NGOs have started to organize community meetings, teaching the small-holder farmers their rights and how the institutional framework works. Additionally, fishermen who are not able to pursue their fishing business anymore are taught simple farming techniques during other community meetings. Thus, political marginalization is not yet viewed as a stressor to intensify existing tensions between the primary conflict actors but might be considered potential conflict drivers in the mid-term.

Economic actors are characterized by fishermen and farmers as conflict stressors because they increase the competitiveness between the farmers and fishermen over land at the lake and therewith water access directly. Between the two research visits, economic companies enhanced their engagements around the lake, resulting in the closing of one landing site (Bugolo). Hence, the fishing community had to leave their small village. During an interview with the residents of another fishing village called Buzi Bazi, the participants mentioned that the fishermen were relocated into the wider surrounding of Lake Wamala. Furthermore, they were unanimous in the view that demonstrations, blockades or any other form of violent behaviour against the companies' acquisition of the land results in the retention of the financial compensations or even leads to detentions. Thus, protests are not considered promising means to remain on the land. Rather, the fishermen prefer to take the money which improves their situation at least for a very short period of time. However, financial grants accelerate the conflictual tensions between farmers and fishermen. Fishermen use the money to either buy small plots of land to fence off land which small-holder farmers used for decades or pay women for sexual favours. This results in and increases inter-communal clashes between the fishermen and the farmers. Furthermore, farmers also reported that fishermen use the money to buy alcohol. In return, the consumption of alcohol increases the aggressiveness of the fishermen against the other community members, and thus, intensifies the existing social tensions.

Currently, the conflict dynamics between small-holder farmers and fishermen are accelerating. Economic and political inequalities aggravate the risk of conflict as an ineffective institutional framework, lack of formal regulations and a patronage network prevents law enforcement or the empowerment of administrative and political structures on a district level. Furthermore, the national economic agenda is driven by the president's aim to invest in infrastructure, road constructions and other macro-economic projects. Thus, the needs and capabilities of the local population are not taken into consideration. Lastly, the farmer-fishermen conflict is triggered by climate change. The gradual changes in the weather conditions, which are intensified by environmental destructions and unsustainable farming techniques, make farming more unpredictable. This can result in the drying of crops (during longer heatwaves) or the flooding of agricultural lands (during heavy rainfalls).

Additionally, the conflict situation is triggered by spoilers and shadow actors. District authorities can be considered spoilers. As a result of President Museveni's attempts to continuously extend his duration in office, underlying tensions within the population built up towards him (Employee German Foreign Office 2018; The Daily Nation 2018). To manifest this power and to limit political opposition, he builds his power on a patronage network which is constantly increasing. As a result, more and more districts, sub-districts and catchment zones are set-up to create employment opportunities for Museveni friendly people (Employee

German Foreign Office 2018). On the one hand, these supporters work within the institutional and legislative framework which is provided by the president and the national authorities. As such, they are supporting the economic development agenda despite its negative impacts on the local population. On the other hand, the president's patronage network puts a lot of strain on the county's financial resources which in contrast negatively affects the monetary situation of the districts and catchment authorities.

The research findings point out that international economic actors can be considered shadow actors. Interviewees reported that Indian and Chinese businesses are the masterminds who operate through national businesses. Thus, Ugandan companies acquire land, set up the economic sites and supervise the operations. However, Chinese and Indian businessmen dictate the national companies on what to do on the ground and influence national authorities in their economic decision-making.

Against this background, conflicts between small-holder farmers and fishermen at Lake Wamala frequently revolve around issues of contested land use and access to water at Lake Wamala's shorelines directly. Furthermore, the conflict dynamics are also driven by relocations and the adaptation to a new living and working environment. Increasingly, these conflicts are triggered or exacerbated by the inadequate implementation of the institutional framework and the proceeding economic operations. Climate change puts additional stress on the resources and the farming and fishing practices. Lastly, drought-induced movement of pastoralists to resource-rich areas might become a further source of tensions and conflict between the communities in the mid to long term. However, this remains subject to speculation, as pastoralists hardly pass by Lake Wamala nowadays.

5.3.4 Conflict Mapping

Figure 5.7 shows the mapping analysis of the small-holder farmer and fishermen conflict based on the observations and interviews conducted during the research visits. The conflict between the fishermen and the farmers is the main issue and the basis for violent and non-violent responses to the conflict dynamics. The situation is accelerated through the interaction with villagers over land and water pumping stations while also living in the wider surroundings of Lake Wamala. Whereas the stakeholders within the dashed circle are in conflict with each other, the secondary actors (Government of Uganda and the District Government of Mityana) negatively influence the conflict context. Economic stakeholders predominantly influence the governmental stakeholders' economic activities, who, in turn, implement those guidelines using a top-down decision-making approach. External actors, such as NGOs, international organizations and community-based organizations, influence the conflict situation indirectly through their connection to the national ministries via financial aid or development assistance. Shadow actors and spoilers are also included in the map. The discussion of the main conflict actors, including their positions, interests and needs in matters related to the conflict issue revealed that the underlying needs of both primary conflict parties are almost similar.

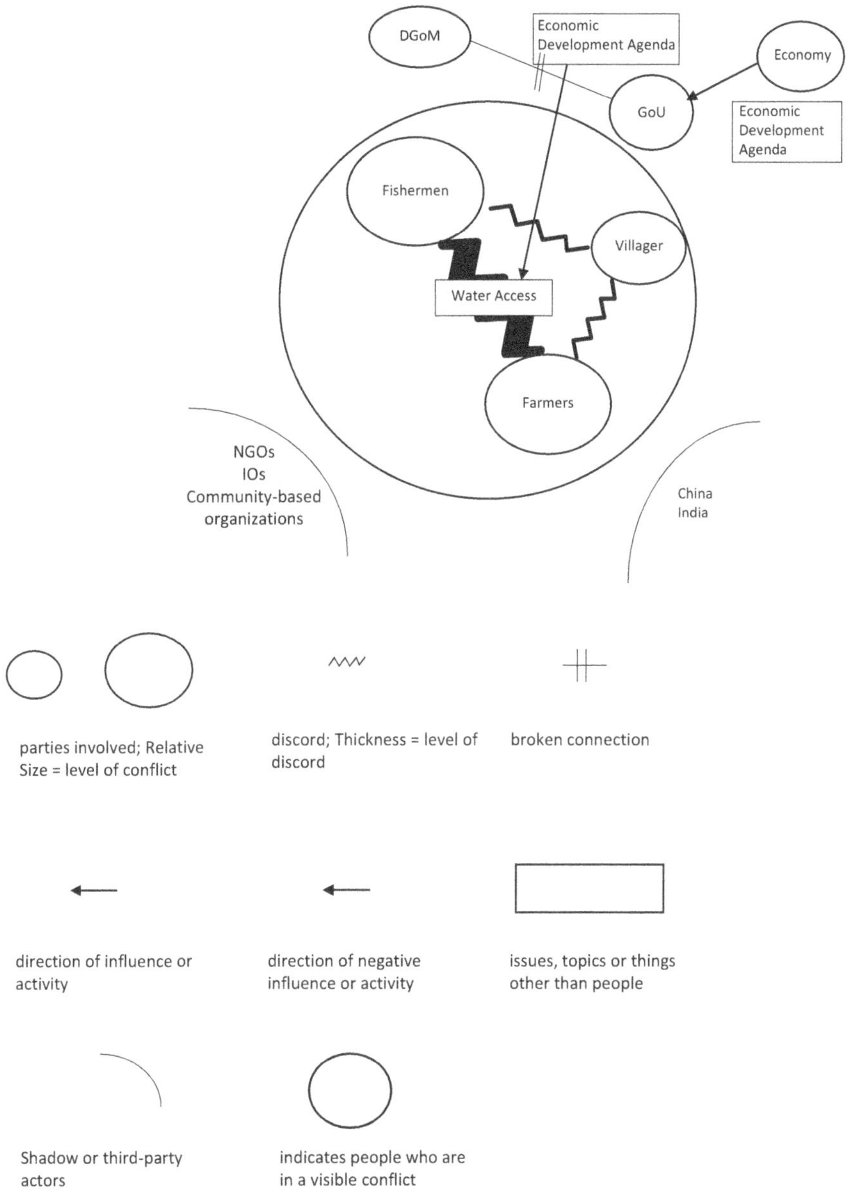

FIGURE 5.7 Conflict Map, Fishermen-Farmer Conflict, Summer 2021

(Source: The Author 2020)

However, they are overshadowed by divergent positions and interests which determine the actors' capacities to exert power over the other primary conflict actors. The conflict actor description disclosed that the conflict dynamics between the primary and tertiary actors is driven by the secondary conflict actor's interests and political and economic strategies. Furthermore, in the case of Lake Naivasha, the currently existing conflict dynamics are the result of both historically grown grievances between the small-holder farmers and pastoralists and the accumulated horizontal political and economic inequalities. In the case of Lake Wamala, the conflict is caused by the primary actor's low capacity and ability to adapt to the changing environment. The conflict analysis illustrated that the topics associated with the broken connection between the various conflict actors are different. This, furthermore, results in latent and visible conflict dynamics. Thus, whereas the conflict between the secondary and external actors remains dormant, the conflict dynamics between the primary and tertiary actors are erupting. The last past of this chapter discusses which response tools the primary and tertiary actors use regarding the previously mentioned conflict dynamics.

5.4 Response Tools to the Conflict Dynamics

Figure 5.8 combines an overview of the tools the local resource users use to respond to the conflict dynamics across both research sites. The economic investments put high pressures on the accessibility of the water and land resources, water provision and the general development of both lake basins. This, in return, puts pressure on the local resource users to adapt to the changing environment. Due to their low adaptive capacity, local resource users turn to non-violent and violent means to limit their high sensitivity to the changes experienced in matters related to water access especially.

Local resource users identified cattle raiding and the blockade of access routes to the lake or water pumping stations as the most commonly used response tools. The results highlight that non-violent and violent response tools are almost balanced nowadays. The most striking result to emerge from the interviews and the performed data analysis is that deathly encounters between the single local resource user groups and individuals is increasing.

This is, furthermore, reflected in the responses of the interviewees who reported on higher levels of insecurity in the informal settlements around Lake Naivasha especially. The distribution of non-violent and violent response tools across both research sites shows that attacks take place mostly in the evenings or at night. Resulting thereof, respondents indicated that small-holder farmers, villagers and fishermen have put up night watches. Small-holder farmers fence off their land to avoid crop raiding or the destruction of their agricultural land, while fishermen protect their fishing equipment, boats and fishing. The majority of the respondents in the informal settlements at Lake Naivasha and within the community settlements around Lake Wamala argued that night guards protect the access routes to the lake, the pumping stations or the few water pipes, if there are any, in the informal settlements. A few respondents highlighted that these night watches are necessary to avoid either a diversion of the water pipes to individual residents' houses or the demolition of the water stations. Generally, the levels of insecurity and suspicion between the local resource users was expressed in the interviews and

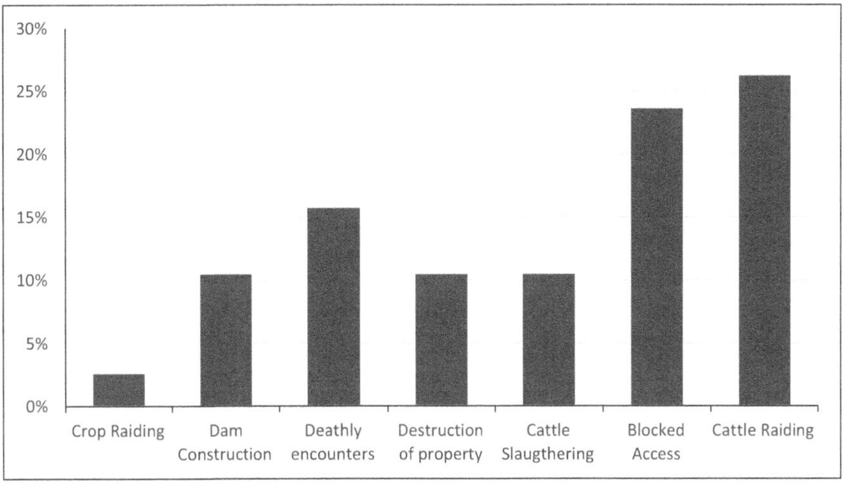

FIGURE 5.8 Response Tools to the Conflict Dynamics

(*Source:* The Author 2020)

is reflected in the response tools the four user groups use against each other (see Figure 5.8).

At Lake Wamala, small-holder farmers mentioned that cases of food theft and cattle raiding happen especially during the dry season. The motivation of the farmers is to either consume the agricultural products or sell it at the local market. Similarly, cattle are either sold for a small fee or used for domestic purposes. Moreover, farmers mentioned that the destruction of fences around agricultural land or housing premises are increasing. Around Lake Naivasha, small-holder farmers indicated that denied access, cattle raiding and cattle slaughtering as well as deadly encounters are the most commonly used response tools. Around Lake Wamala, pastoralists hardly pass by. However, members of Kikandwa community mentioned that conflicts between pastoralists and the other resource users erupt quickly when pastoralists try to access the lake and its water and land resources. The most commonly used form of non-violent behaviour is the destruction of property as pastoralists push their cattle through agricultural areas down to the lake's shorelines. Nevertheless, the percentages of other reported incidents are lower than in the case of Lake Naivasha. Around Lake Naivasha, both cattle raiding and cattle slaughtering, as well as deadly encounters, have the highest values. Cattle raiding is either used to re-stock the pastoralist's own herd or as a means to pay dowry. The latter especially happens in traditional community settings. These numbers are only exceeded by the values mentioned for denied and blocked access. The high values for blocked access routes can be explained by the villagers' and small-holder farmers' wish to protect water holes, water stations and farmland. As already mentioned, there is little conflict between the fishermen. In the case of Lake Wamala, fishermen destroy farmland, block access or steal crops and agricultural products from small-holder farmers. Around Lake Naivasha, cattle slaughtering and cattle raiding are regularly used tools against pastoralists. Furthermore, the blockade of the access routes to the landing sites, as well as deadly encounters by and between the different user groups have increased over the past years. Lastly, villagers and employees in the flower farms or hotels reported that they have also been targeted by non-violent and violent response tools. At Lake Wamala, the destruction of property and cattle raiding have been mentioned the most. Around Lake Naivasha, cattle raiding, the blockade of water stations and access routes to the landing sites and deadly encounters are frequent.

Generally, Lake Naivasha is more affected by non-violent and violent behaviour than Lake Wamala (see Figure 5.9). The figure illustrates the conflict intensities for the two lake sites separated for water conflict and general conflict response tools. During the interviews, the participants were asked about their main motivations for applying non-violent and violent response tools. At both lake sites, the underlying motivation of conflict is less driven by water per se. Instead, stakeholders mentioned other reasons for the conflict dynamics and, therewith, this results in the high percentages for general conflict intensity. The higher numbers for Lake Naivasha may be explained by the fact that the conflict dynamics are shaped by historically grown grievances and ethnic affiliations. Another possible explanation for this is that there are more people living in Lake Naivasha basin, thus the demand for the diminishing resources and space around the four remaining landing sites increases. What stand out in the figures are the higher values for water conflict at Lake Wamala compared to Lake Naivasha. This difference compared to the data for general conflict can be explained in part by, firstly, the relatively new situation

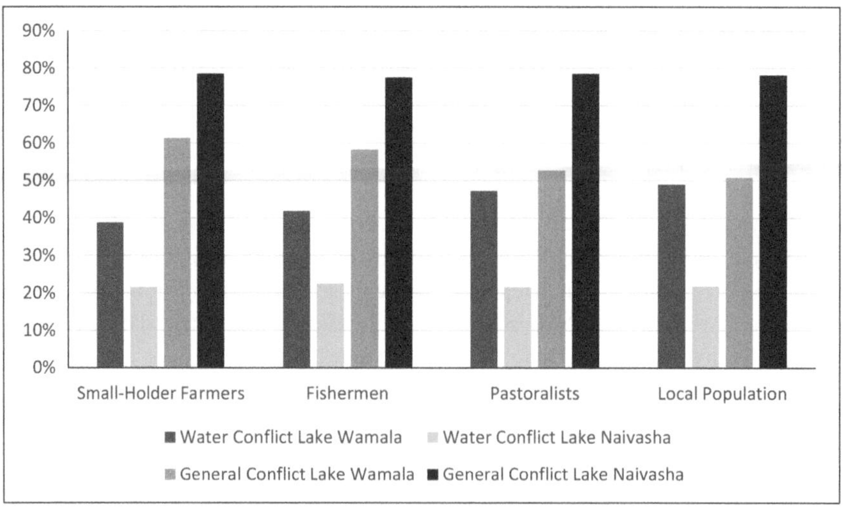

FIGURE 5.9 Conflict Intensity at Lake Naivasha and Lake Wamala

(Source: The Author 2020)

for the local resource users to adapt to the changing situation. Resulting thereof, the issue of water access, especially the lack thereof, superposes other motivations for conflict. Secondly, local resource users have been used to adapt to the living conditions without external assistance, or even expectations towards them. Thus, for a long time, political, economic and social changes happened without notice of the local resource users and, therefore, have never been seen as a reason for conflict. Accordingly, current conflict dynamics, which are also slightly driven by demands for health centres, better road infrastructure or education facilities emerged with the arrival of the economic companies. The observed increase in general non-violent and violent behaviour could be attributed to the increasing economic investments undertaken. The numbers are still lower than at Lake Naivasha. Nevertheless, the frequency of non-violent and violent response tools to the changing environment is rising.

During the interviews and focus group discussions, the interviewees also mentioned other response tools to the conflict dynamics. Among others, verbal violence, conflicts between human beings and wildlife, adultery, drunkenness and domestic violence have been named by the interview and focus group participants. What was surprising during the research visits was the openness of the local resource users about the response tools used vis-a-vis the other resource users. Furthermore, one unanticipated finding was the high intensity of violent behaviour. When asked about the motivations for non-violent and violent behaviour, the majority of the respondents highlighted that violence is applied to reduce the number of residents and local resource user groups living and sharing the space around the lake. They indicated further that by aggravating the living situation for any group, these groups might move away from the basin in search of a new and securer living environment. Resulting thereof, the amount of people straining the landing sites and using the few land and water resources available decreases. A few respondents expressed the view that violent response tools are used to raise the national and international communities' awareness about their living situation. Between the two research visits, the levels of non-violent and violent response tools increased. However, the conflict dynamics are continuing without much notice from the other actor groups.

5.5 Brief Summary

The aim was to assess and compare the conflict dynamics at Lake Naivasha and Lake Wamala with an emphasis on the temporal and local resolution, the motivation and the response tools used. Firstly, a systematic study of the actors' positions, interests and needs, root causes and conflict profile were carried out. Secondly, the conflict analysis captured the conflict dynamics contributing to violent conflict and the multidimensionality of the conflict patterns, including political, social, economic or environmental fields of policy. Thirdly, the conflict map highlighted the relationship between the directly and indirectly involved conflict actors and the main conflict issue. The conflict assessment for the research sites revealed different motivations for the currently existing conflict dynamics. Whereas the current conflict dynamics are accelerated by historically grown grievances and ethnic affiliations at Lake Naivasha, at Lake Wamala, the conflict dynamics are the result of the relatively recently started economic investments and relocations of the fishermen and small-holder farmers. Furthermore, political kinship, economic investments and environmental destructions intensified the conflict situation at Lake Naivasha because the government has failed to provide the local stakeholders with basic services, including education and health facilities or water infrastructure. This has led the communities to demonstrate against the companies and the government for better working conditions, the improvement of water provision or against environmental destructions. However, dismissals or reduced water provision drove the communities to transfer their unmet expectations for development, employment or resource access away from the government and the companies onto other local resource users. The new living environment and the economic investments undertaken in conjunction with the ongoing process of decentralization have resulted in emerging conflict dynamics between the small-holder farmers, fishermen and villagers at Lake Wamala.

Both the temporal resolution and the conflict motivations resulted in non-violent and violent response tools to increase the stakeholders' adaptive capacity to external changes. While non-violent behaviour, including the blockade of access routes, destruction of fences or crop and cattle raiding was a commonly used tool at the beginning, violent response tools are often used instead. In general, the conflict intensity is reported to be higher at Lake Naivasha than at Lake Wamala. However, the conflict dynamics around Lake Wamala are rising. There is a likely scenario that economic investments and political kinship will exacerbate existing marginalization and discrimination will fuel more conflict. Lastly, the conflict assessment revealed that the conflict data obtained from ACLED or SCAD do not capture all conflict dynamics at Lake Naivasha and Lake Wamala. Especially latent conflict dynamics and conflicts between local resource users are not captured. Resulting thereof, the lake sites seem calm on the surface; however, there are some serious underlying problems for peace in the area.

Notes

1 Spoilers are defined as individuals or groups that actively seek to hinder, delay or undermine conflict settlement approaches (see, e.g. Newman and Richmond 2006).
2 For more detailed information see also, for example, Bond (2014) or Yurco (2017).
3 During the interview, it became apparent that the not only word 'tribe' included, for the interviewee, an ethnic affiliation, but the member of LNRUA also used the word 'tribe' for other groups of people who are residing in the basin nowadays, including economic stakeholders, tourists or NGOs and political actors.
4 The violent events after the general elections in 2008 and 2016 are exceptions. During this time, the government, as well as security forces, has been involved in violent conflict settlement and resolution. Consequently, the unusual high amount of fatalities in 2008 have to be interpreted with caution as those are less the result of water-related conflicts and more the consequence of the violent post-election upheavals.

5 It needs to be mentioned that inter-ethnic rivalries do not shape the conflict profile to a large extent anymore. However, interviewees reported that historical grievances accelerate existing and upcoming conflict dynamics.

6 Majority of the interviews were anonymised (see Chapter 2). Therefore, resident refers to a person who is living at Lake Naivasha.

7 'Government of Kenya' and 'County Government of Nakuru' serve as umbrella terms for ministries and departments on the national and sub-national decision-making level to avoid an overload of the conflict map.

8 The higher death rates in 2005 and 2006 resulted from the residents' demonstrations against President Museveni's intention to run again in the 2006 presidential election. Years 2010 and 2011 have also been understood against a different background. The government intended to sell parts of Uganda's largest forest (Mabira forest) to Indian investors who aimed to convert the forest for a better economic utilization, in particular for intensive rice growing. The partly strong demonstrations have been dissolved by police and security forces resulting in higher number of deaths.

9 Fishermen and small-holder farmers around Lake Wamala belonged all to the same tribe and have been members of Baganda Kingdom. As such, across the interviews, ethnic rivalries have not been mentioned as potential conflict drivers.

Bibliography

Adem, T. A., Ember, C. R., Skoggard, I., Jones, E. C. and Faas, A. J. (2012): Dangerous geography: Spatial distribution of livestock raiding in northwestern Kenya. *Ethnology* 51 (1), 1–29.

Armed Conflict Location and Event Dataset (2021): *Armed Conflict Location & Event Project. Bringing Clarity to Crisis*. At: www.acleddata.com/ [retrieved last 10 November 2021].

Balama, J. (2019, 15. August): Interview by Renner, J. Nairboi, Kenya.

BBC (2018): *Kenyatta Wants Legacy to Be Fight Against Corruption*. At: www.bbc.com/news/av/world-africa-45343956/kenyatta-wants-legacy-to-be-fight-against-corruption [retrieved last 31 March 2020].

Bond, J. (2014): A holistic approach to natural resource conflict: The case of Laikpia County, Kenya. *Journal of Rural Studies* 34, 117–127.

Community Member Karagita (2019, 20. August): Focus Group Discussion by Renner, J. Karagita Landing Site, Kenya.

Daily Nation (2018): *Uganda's Museveni Set for 'Life Presidency' After Court Ruling*. At: www.nation.co.ke/news/africa/Uganda-Yoweri-Museveni-life-presidency/1066-4684346-13er6ii/index.html [retrieved last 31 March 2020].

Ember, C. R., Abate, A. T., Skoggard, I. and Jones, E. C. (2012): Livestock raiding and rainfall variability in northwestern Kenya. *Civil Wars* 14 (2), 159–181.

Employee German Foreign Office (2018, 17 August): Interview by Renner, J. Kampala, Uganda.

Fisher, S., Ibrahim Abdi, D., Ludin, J., Smith, R., Williams, S. and Williams, S. (2000): *Working with Conflict: Skills and Strategies for Action*. Zed Books. New York.

Galtung, J. (1996): *Peace by Peaceful Means: Peace and Conflict, Development and Civilisation*. PRIO. Oslo.

Goldsmith, P. (2013): The future of pastoralist conflict in the horn of Africa. In: Cattley, A., Lind, J. and Scoones, I. (Eds.): *Pastoralism and Development in Africa: Dynamic Change at the Margins*. Routledge. Abingdon.

Herbert, S. (2017): *Conflict Analysis: Topic Guide*. GSDRC, University of Birmingham. Birmingham.

Kapila, S. (2018, 10 August): Interview by Renner, J. Naivasha, Kenya.

Kundu, R., Aura, C. M., Muchiri, M., Njiru, J. M. and Ojuok, J. E. (2010): Difficulties of fishing at Lake Naivasha, Kenya: Is community participation in management the solution? *Lakes & Reservoirs: Research and Management* 15 (1), 15–23.

Newman, E. and Richmond, O. (2006): Peace building and spoilers. *Conflict Security and Development*, 6 (1), 101–10.

Poolman, M., Munamati, M. and Senzanje, A. (2009): *Stakeholder and Conflict Analysis. Small Reservoirs Toolkit.* At: http://www.smallreservoirs.org/full/toolkit/docs/I%20 02%20Stakeholder%20and%20Conflict%20Analysis_MLA.pdf [retrieved last 2 December 2021].

Ramsbotham, O., Woodhouse, T. and Miall, H. (2011): *Contemporary Conflict Resolution.* Third Edition. Polity. Cambridge.

Ropers, N. (2008): *Systemic Conflict Transformation: Reflections on the Conflict and Peace Process in Sri Lanka.* Berghof Hanndbook Dialouge No. 6. Berlin.

Schilling, J., Akuno, M., Scheffran, J. and Weinzierl, T. (2014): On raids and relations: Climate change, pastoral conflict and adaptation in north-western Kenya. In: Bronkhorst, S. and Bob, U. (Eds.): *Climate Change and Conflict: Where to for Conflict Sensitive. Climate Adaptation in Africa?* Berliner Wissenschaftsverlag. Berlin.

Schilling, J., Locham, R., Weinzierl, T., Vivekananda, J. and Scheffran, J. (2015): The nexus of oil, conflict, and climate change vulnerability of pastoral communities in northwest Kenya. *Earth System Dynamics* 6 (2), 703–717.

Schilling, J., Opiyo, F. and Scheffran, J. (2012): Raiding pastoral livelihoods: Motives and effects of violent conflict in north-western Kenya. *Pastoralism* 25 (2), 1–16.

Shettima, A. G. and Tar, U. A. (2008): Farmer-pastoralist conflict in West Africa: Exploring the causes and consequences. *Information, Society and Justice Journal* 1 (2), 163–184.

Social Conflict Analysis Database (2021): *Social Conflict Analysis Database.* At: www.strauss center.org/scad.html [retrieved last 10 May 2021].

Vivekananda, J. (2015): *Peace Audit Kenya.* International Alert. London.

Yurco, K. (2017): Herders and herdsmen: The remaking of pastoral livelihoods in Laikpia, Kenya. *Pastoralism* 7 (15), 1–11.

6

FROM CONFLICT TO CO-OPERATION

Rethinking Water as a Generator of Peace

6.1 Drivers of Water-Related Conflicts

As indicated previously, the conflicts between the local resource users frequently revolve around issues of water access and contested land use. The analysis has shown that these conflicts are triggered by the institutional framework and the economic development agendas (see Table 6.1). The results, furthermore, show that farmer-herder and farmer-fishermen conflicts occur independently from the seasonal changes.

When asked about the main causes of conflict (see Figure 6.1), community members expressed the view that the economic undertakings are not the problem. Instead, they stressed that the national political system and its main stakeholders are the main issues (30 per cent at Lake Wamala and 29 per cent at Lake Naivasha). Thus, the insufficient implementation of the policy rights, as laid out in the countries' constitutions, negatively affects the situation between the local-level stakeholders.

At Lake Naivasha, a few community members blame the government in Nairobi for the situation in the basin. They highlight that the county government, and, therefore, the local population, does not benefit from the revenues. More recently, they voiced their displeasure about the county government representative's inactivity to demand law enforcement and decision-making as laid out in the constitution. They argue that high levels of corruption and the promise of economic and political benefits is limiting county involvement in local affairs or even working against the local-level actors. Thus, they feel politically marginalized and ignored by both county and national government representatives. Similarly at Lake Wamala, people do not feel that they have the power to change anything. Moreover, they seem to be unwilling to change anything as they only want to follow national objectives.

DOI: 10.4324/9781003271291-6

TABLE 6.1 Causes of Conflict and Conflict Motivations and Capabilities of Local-Level Actors at Lake Naivasha and Lake Wamala

Causes of Conflict	*Conflict Motives*		*Conflict Capabilities*
	Lake Naivasha	*Lake Wamala*	
1. Political power contestations	Improved water access and water infrastructure	Defending of territory	Experience in attacks
2. Territorial control over key resources	Improvement in education and health infrastructure	Improved water and land access	Knowledge of are and the communities
3. Unclear land rights and governmental land policies	Defending of territory	Financial compensations	Weapons and ammunition (partly)
4. Unmet expectations	Payment of dowry	Hunger/Poverty	
5. Ethnic polarization, historical grown rivalries and revenge	Avoiding of population increase		
6. Economic marginalization			

(*Source*: The Author 2020)

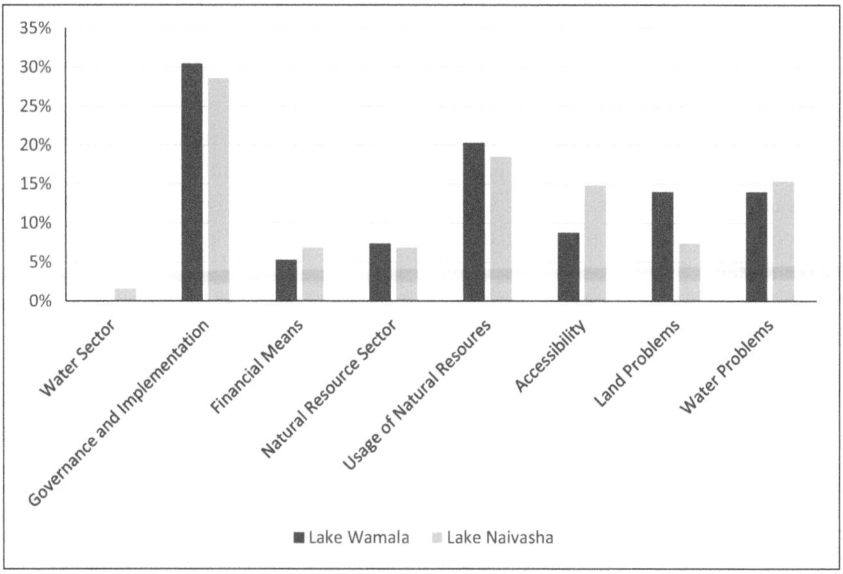

FIGURE 6.1 Causes of Water-Related Conflicts at Lake Naivasha and Lake Wamala

(*Source:* The Author 2020)

Thus, securing access to water and land and other necessary resources to sustain a living seems to be a crucial prerequisite to the local resource users to cope with difficult social and economic drivers in the lake basins (15 per cent for Lake Naivasha and 9 per cent for Lake Wamala). The economic investments undertaken led to changes in land use and, hence, obtaining territorial control over key resources was identified as an important conflict cause. Whereas pastoralists aim to secure their mobility, farmers want to expand their cultures. Due to the relocations of fishermen into the wider surrounding of Lake Wamala, fishermen obtain farmland. Because small-holder farmers, pastoralists, villagers and fishermen utilize the same land, conflicts erupt easily (14 per cent for Lake Wamala and 7 per cent for Lake Naivasha). The former is intensified through unclear land tenure and governmental land policies. Traditionally, land along the South and West of Lake Naivasha and around Lake Wamala is communal land, which means that there is no individual ownership of the land.

This, however, has created two problems. Firstly, farmers, pastoralists and villagers have issues concerning land use (crop cultivation vs. grazing vs. construction of houses). Secondly, the government lends communal land to economic companies to cultivate roses, produce geothermal energy, to do extensive rice growing or to explore the area for other economic undertakings. The communities around Lake Naivasha perceive these economic actors as stressors who have taken their farming and grazing land without improving their living situation as promised, including, well-paid labour opportunities, improved access to water and other infrastructural investments. Hence, most land around Lake Naivasha is private land which is inhabited by families or economic stakeholders currently. Local stakeholders are left to the four remaining sites and to the land inside the informal settlements. Moreover, the communities around Lake Naivasha did not receive any compensation from the Kenyan government after they rented out the land to economic stakeholders. Once the businesses started to operate and to generate incomes, the Constitution of Kenya requires an equitable revenue sharing from nationally earned revenues between the national government (70 per cent), the county government (30 per cent) and the local communities (10 per cent) (Government of Kenya 2010). However, the issue of revenue sharing is causing tensions between the government and the counties, thus the local communities do not receive their revenue share. Similarly, at Lake Wamala, the communities at the three remaining landing sites perceive the economic actors as stressors who come in and take their land without prior consultation. As mentioned, the fishermen at the landing sites received small financial compensations from the economic investors. Nevertheless, the community members who are living in the wider surroundings of the lake do not receive any financial compensation and, therefore, fear the newly arrived fishermen more than the economic actors. Consequently, they fence off their land or agricultural sites to defend their territory.

At Lake Wamala, the land issue is even more intensified through the misunderstanding about mailo land. The misunderstanding is created as mailo tenure generates dual ownership over the same piece of land (Uganda Land Act 1998; The Observer 2019). Mailo landowners have the same rights as freehold land owners; however, according to section 3 (4) of the Ugandan Land Act, they have to respect 'the rights of lawful and bonafide occupants[1] and Kibanja holders[2] to occupy and live on the land' (Uganda Land Act 1998). Moreover, the Kabaka's[3] land also falls into the category of mailo land. As a result, the Kabaka's land cannot be sold entirely as some land belongs to the Kabaka personally, while other parts of the land can accommodate Kibanja holders. Additionally, some portions of the Kabaka's land are used for spiritual activities or for cultural rituals. Resulting thereof, there is a lot of misunderstanding about interests and rights to the land around Lake Wamala and the land ownership, respectively. Moreover, the four different local resource user groups disagree about the individual set boundaries and land demarcations.

Unmet expectations are the next critical conflict issue. The communities around the lake basins articulated that they have been expecting improvements in both educational, health and road infrastructure, as well as improvements regarding water access. Furthermore, the residents of the informal settlements and incomers from East and North Kenya expressed their wish for employment opportunities in the economic sites around Lake Naivasha. According to the county representatives and a flower farm representative, almost all jobs in the flower farms are meant for local people. They highlighted that they adhere to international working standards. Thus, community members receive employment benefits, including free transportation from Naivasha town to the flower farm, a 7.5-hour working day and a regular income. However, they also acknowledged that the local people are not qualified for senior positions due to their low level of education. According to villagers and employees, workers in the flower farms are exposed to inhuman working conditions. There are different types of work, for example, flower picking, spraying and cutting or security and cleaning. Employees reported that they are exposed to the chemicals directly as they have to harvest and treat the cut flowers without protection. Workers reported that they earn between 50 and 100 USD per month. This is not enough to sustain a living for them and their families (Vivekananda 2015; BBC 2017). Moreover, employees mentioned that they have to work between 10 and 11 hours per day. Lastly, most of the workers are female. Some of them reported on gender-based violence and sexual exploitations of senior workers to either increase their monthly salary or get promotions.

When asked about employment in the geothermal sites, villagers argued that they will not be able to work in these sites as they do not meet the qualifications. The most senior position a local resident might get is as a security guard. Given that educational facilities are still rare, local residents do not expect any improvement in the short to medium term. At Lake Wamala, people do not expect any employment opportunities in the newly constructed economic sites. According to villagers, the companies come in with their own manpower. Hence, there is very low demand for unskilled labour. Instead, they suggested that these companies should improve the road infrastructure and education opportunities around the lake and in its surroundings. The two field visits revealed that investments in road and education infrastructures have not started yet. The roads around the lake are in really bad conditions, and especially during the rainy season, people are unable to access the lake's landing sites or access Mityana town and the main road. Because villagers are unable to sell their fruits and fish to traders from Kampala, the prices drop as they cannot be sold in the major cities. Lastly, education around the lake remains really low as schools are either not operating or not existing. Conversely, this means, that unemployment rates around the lake remain very high.

Ethnic polarization, historically grown grievances and revenge are the next conflict driver, especially around Lake Naivasha. After independence, Jomo Kenyatta placed members of his own ethnic group (Kikuyu) across the country to both gain access and control over the land. Consequently, tribes created artificial borders which are often questioned by farmers, pastoralists and communities. Cross-border movements of farmers and pastoralists do not affect the inter-state relationships on a national level even though the national land rights are working against the national policy structures which have changed in the 1980s and revised in 2012 again. Thus, ethnic polarization and historically grown rivalries are a sub-national problem between the local-level actors. The motive of revenge intensifies the conflictual situation between them.

Economic marginalization was less mentioned as a driver of the conflicts, although it is present and is influencing the local conflict dynamics (7 per cent for Lake Naivasha and 5 per cent for Lake Wamala). The economic sites divert water which was originally used for agriculture. Thus, around Lake Naivasha and Lake Wamala, water becomes commercialized and therefore, allows the economic companies, political actors and governmental loyal people to manage water and dams at their own discretion and to their own prices. Community members expressed hope that the economic sites or national water providers would give them water. To a limited extent, National Water and Sewage Cooperation (Uganda) and Naivasha Water and Sewage Cooperation (Kenya) drill boreholes or set up water tanks along the road. Additionally, to the researcher's knowledge, economic companies provide piped water to the informal settlements of Karagita and Kameri. This, however, does not mean that the entire settlements have access to piped water. Generally, there are a few pipes which supply a few water points throughout the entire area.

Community members and residents in the informal settlements reported that water was provided only at the beginning on a regular basis (15 per cent for Lake Naivasha and 14 per cent for Lake Wamala). Moreover, they emphasized that neither the economic companies nor the water suppliers met their expectations for water. At both lake sites, local resource users highlighted that the bore holes and water stations are 'far away' from their settlements and, moreover, one water source cannot provide enough water to support livestock, the agricultural sites and the communities. Even though the water suppliers were not available for any interviews, observations revealed that clean and fresh water can be purchased. However, the communities are unable to buy water from the water suppliers. Thus, there is nothing else they can do but to use the water from the lakes. The demand for improvements in health and education facilities has not been met. Some efforts have been made to improve health care and education facilities. Nevertheless, schools remain empty and health centres are out of reach of the communities. Resulting thereof, the level of education remains low and, accordingly, unemployment rates remain high. Furthermore, many community members are in poor health because of bad human sanitation,[4] water-borne diseases and high HIV quotas. Taken together, these factors undermine the economic and social situation of the local resource users and are likely to become more prominent conflict drivers in the future.

Even though both lake basins have not been affected tremendously by climate change yet, the economic investments at both lake sites are going against the climate goals of Kenya and Uganda. There is a risk that the economic undertakings intensify the basin's exposure to climate change. Thus, increased droughts or heavy rainfalls can force small-holder farmers and pastoralists to change their mobility routes and agricultural sites and aggravate their competition for water and land among each other and with other communities.

The majority of the community members and local resource users expressed the view that the security situation is worsening since the economic investments have started. When asked about the drivers for increased insecurity levels, they noted that the conflict motives centre around access to water and land, as well as the defence of water infrastructure or land ownership. They furthermore highlighted that neither the economic companies nor the political stakeholders want to mitigate or respond to the conflict drivers between the local resource users. A landing site overseer indicated that political and economic stakeholders remain ignorant to the local people's concerns and vulnerability to the economic investments undertaken. This perception was supported by the responses received from the conducted interviews with political, economic and international stakeholders. After introducing the main focus of the research, some interviewees asked why Lake Wamala or Lake Naivasha were chosen as research sites as they are not aware of any ongoing conflicts in these areas. In the case of Lake Wamala, a few interviewees even asked where the lake is located as they have never come across it previously.

Due to the insecurity arising from both the intensified conflict dynamics and the economic situation, most of the agricultural land and the water access points are either destroyed or over-utilized, implying that all groups of local stakeholders are confined to smaller and scuffed land. Thereby, they contribute to the degradation of the farm- and range-land and negatively affect the water quality and water availability. The depletion of the available water and land resources during both seasons have negative ecological impacts as the farmland loses productivity due to encroachments and the use of chemicals or pesticides. Resulting thereof, undesirable species invade the lake basins and replace key flora and fauna (Donald et al. 2006). To cope with the situation, local resource users opt to plant both fast-growing and highly utilizable species (e.g. Eucalyptus trees), as well as choose to use fertilizers and chemicals to boost the crop productivity. This, however, results in undergrowth suppression, an increase in soil nutrient or water depletion (Zerga

2015). When asked about the effects of foreign species on the lake basin's biodiversity, community members indicated that some of them are aware of the consequences. Yet, they also argued that population pressure, the high demand for water, agricultural products or timber overshadow the need to plant trees or to conserve the basins resources in ideal site.

6.2 The Nexus of Water, Conflict, Vulnerability and Governance

Figure 6.2 summarizes the key findings. The analysis showed that key to understanding the outbreak of water-related conflicts are the institutional and governance

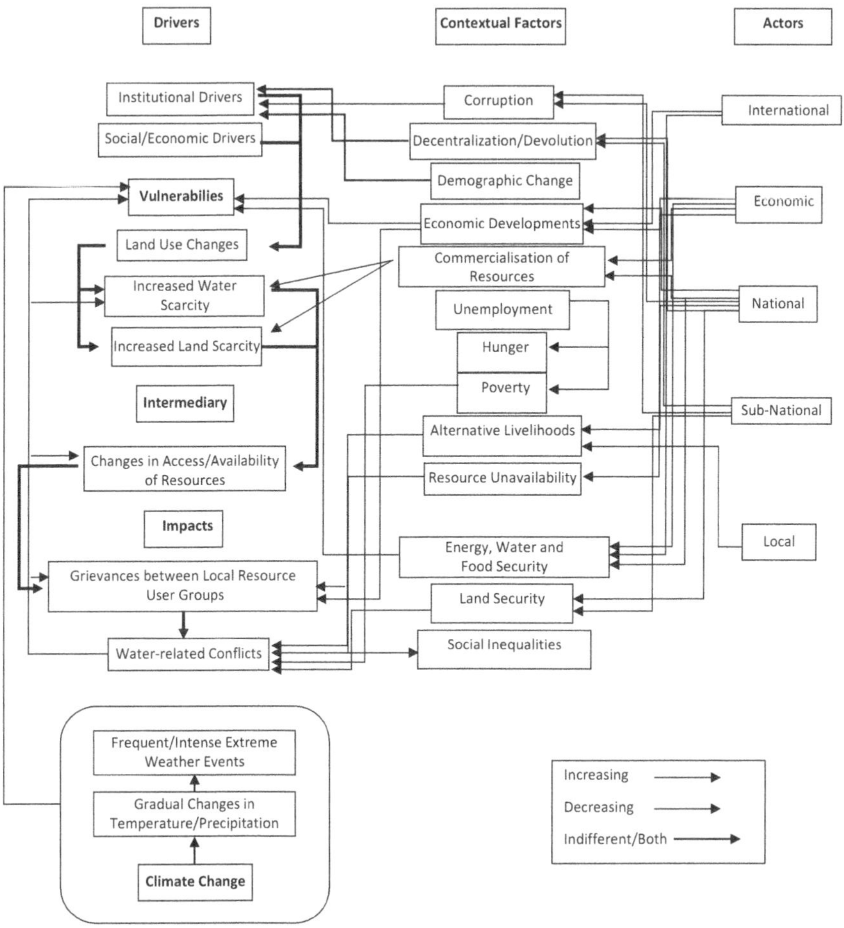

FIGURE 6.2 Schematic Overview of Drivers, Contextual Factors and Actors Influencing Water-Related Conflicts

(Source: The Author 2020)

dynamics. During the past decade, Kenya and Uganda have given attention to change the institutional framework. The change of the institutional context was also linked to the wish to be better integrated into the world market and to become a player in a more globalized world. However, the changing institutional framework and the economic orientation is linked to a number of concerns and controversies over natural resource security and competition, social and economic impacts and inadequate government regulations.

6.2.1 The Role of the Institutional Framework and the Economic Agenda in Shaping Water Shortages

Beside the already-discussed drivers of water-related conflicts, the figure shows that the conflicts have significant direct and indirect effects on the local-level actor's wellbeing. It, furthermore, highlights that the causes of water-related conflicts are less water-driven than institutionally made. Both Kenya and Uganda are increasingly susceptible to national and international economic developments and institutional changes. These investments range from low-level economic developments including agriculture, horticulture and farmland to macro-economic ones including manufacturing, transport and energy. These economic developments create high expectations among the community members, the political stakeholders as well as the international community. The most important expectation to emerge from the analysis is that the community members expected, and to some extent still expect education, infrastructural improvements, employment and water provision. Political stakeholders want to strengthen the economic profile of the countries and strive to become export-oriented economies and middle-income countries within the next two decades. Resulting thereof, natural resources are commercialized and sold and rented to international economic actors. Both Kenya and Uganda are very attractive for international actors for land, as well as to develop cash crops. Generally, international actors benefit from the involvement on the ground over and above the cash as it gives them the opportunity to influence the operation and strategic direction of the economic developments or to discuss their concerns of good governance. These economic developments lead to changes in land use. In a similar way, demographic changes cause land use changes. Additionally, demographic changes increase the pressure on the available water resources.

Both countries' institutional effectiveness is undermined by the national economic agendas. Therefore, the ongoing decentralization process, as well as the performance of devolution which has been finalized since 2018, is unable to fulfil its water governance and general policy formulation objectives as laid out in the constitutions. In addition, high levels of corruption and political kinship undercut the institutional effectiveness. Thus, the county and district governments have failed to provide the local resource users with basic services, which in return increases their vulnerability to resource access.

6.2.2 Effects of the Institutional and Social and Economic Drivers on the Actors' Vulnerabilities to Water Access

Flower farming, the geothermal companies and other economic investments need significant amounts of ground- and lake-water which the local resource users depend on. The population growth increases the pressure on the available water resources. Resulting thereof, the communities at Lake Naivasha and Lake Wamala experience water shortages or water scarcities even though both lake's water levels seem sufficient high from the outside. The economic developments and demographic changes shift the land ownership and, hence, reduce the available and usable land. Aside from the mentioned vulnerabilities, other external factors (high unemployment rates, food security and poverty) worsen the local people's vulnerability. By the time the research was conducted, these externalities were less mentioned and noticeable as conflict drivers. However, they are likely to become a source of conflict or exacerbate the existing tensions between the local resource users. Pollution and unsustainable farming activities (deforestation, use of pesticides or degradation of natural resources) of the local communities and the contamination associated with the economic investments contribute to the lake basin's exposure to environmental and climate change and, thus, increase the local resource user's vulnerability. Climate change leads to gradual changes in temperature and precipitation. Additionally, there are more frequent, intense and extreme weather events. Changing climate leads to decreased water and land availability. Similarly, extreme weather events result in reduced water levels and available land. Together, these factors have the potential to reduce the adaptive capacity of all local resource users.

Increased land scarcity hampers the agricultural production of small-scale farmers and reduces the mobility of pastoralists. The limited availability of water access routes to the lakes interact with a parallel increase in people pushing onto the remaining landing sites and water pumping stations. Because of the previously mentioned factors, freshwater becomes noticeably scarce as an essential resource for the local resource users. Thus, the reduced change in access and availability of natural resources, especially water, leads to grievances between the local resource user groups.

6.2.3 The Role of the Institutional and Social and Economic Drivers in Shaping the Overall Impacts on the Local Resource Users Towards Water Access

The current study found that there are different reasons and roots to the conflict. As such, it would be oversimplified to assume that water-related conflicts at Lake Naivasha and Lake Wamala are environmental conflicts only. The factors associated with the outbreak of water-related conflicts are influenced and shaped by contextual factors and have to be understood against a wider background. The social and economic drivers of both countries are determined by a primary focus on development to regain control over the spiralling debt stock.[5] Institutionally, both countries appear to be largely on track in matters related to decentralization. However, the national governments successfully undermine the effects of shifting national responsibilities to the county and district level. Thus, a proper policy implementation as laid out in the constitution is seen as a clear threat to an effective economic development planning and implementation. This leads to an unnecessary duplication and localization of corruption. The institutional, social and economic drivers increase local resource users' vulnerabilities to externalities. The vulnerabilities are intensified and defined by water unavailability or land limitations. To increase their individual adaptive capacity, farmers, pastoralists, fishermen and villagers are looking for alternative sources to recover and to strengthen the agricultural sector. However, they do not sufficiently take into consideration the sector's vulnerability to environmental and climate-related shocks (i.e. droughts, floods, overfishing, health diseases due to the overuse of chemicals and pesticides). The changes in resource availability are determined by the national development plans for energy, land and water security for the economic sites. Nonetheless, these changes negatively impact water and food security for the local resource users. Ultimately, grievances, tensions and conflicts over water and other resources increase and erupt between the local resource users.

An interesting finding which emerged from the study was that water becomes a target between the economic actors, the countries and the communities. The mismatch between the institutional framework and the economic prioritization of the national decision-makers causes multi-scalar conflicts at the same time.[6] Generally, the conflict between the economic and political stakeholders can be classified as a diplomatic conflict. The conflict about political power and its implementation between the national and sub-national stakeholders is resolved through the idea of strengthening economic developments in the lake basins and a rent-seeking behaviour of both the county and district, and the national representatives. Furthermore, political kinship and political and economic favours settle the institutional and diplomatic conflict. In general, therefore, it seems that the diplomatic conflict revolves around the narrative of economic transformation. Accordingly, the access to water becomes commercialized and is regarded as a question of the actors' relationship and support from the national government on the one hand, and the government's dependency on the stakeholders for economic development on the other hand. For this reason, when economic companies come in, local resource users are displaced.

Figure 6.1 displays the percentages of conflict causes for the eruption of water-related conflicts. The conflict causes are set as the default type which can cause non-violent and violent behaviour between local stakeholders at Lake Naivasha and Lake Wamala. The data has been viewed through the lens of non-parametric statistical tests because a Gaussian distribution cannot be assumed, and the variables are independent. A Pearson chi-square test was conducted to detect a link between two variables, namely the variable of conflict cause and the research sites. The variable of research sites includes the two variants 'Lake Naivasha' and 'Lake

Wamala'. The Pearson chi-square test confirms that there is a medium effect on the significant relationship in the tendencies over the outbreak of a water-related conflict in terms of the main detected conflict causes – 'governance and implementation', 'water and land usage' and 'accessibility' between these individual research sites ($\chi^2(9) = 13$, $p = 0.057$, $\varphi = 0.169$).

Additionally, Spearman's rank correlation coefficient was conducted, which is a non-parametric measure of a rank correlation. It analyses the statistical dependence between the rankings of the two variables. The Spearman correlation shows that there is a positive correlation between the conflict cause and the research site ($r = 0{,}923$). Hence, the Spearman correlation indicates that when local stakeholders experience a decline in resource access as a result of a fragmented institutional framework, the greater is their correspondence to violent behaviour. This highlights that the proper implementation of the institutional framework is the default type for resource access and resource distribution. Hence, it could conceivably be hypothesized that the local stakeholders' water-related conflicts are the result of the localization of the diplomatic conflict between the national- and sub-national-level actors. This localization is possible because of the local actors' high vulnerability to external changes. Thereupon, resentments between and against the locals emerged and led to an increase in the manipulation of the groups against each other. Overall, water-related conflicts between the local resource users are the result of institutional and social and economic drivers.

To sum up, the increasing complexity of the institutional framework and the growing influence of external actors confirm the importance of assessing the complex interaction between political actors, economy and civil society internationally, nationally and sub-nationally. In the case of the two lake sites studied, the mentioned interactions resulted in an environment in which local stakeholders are faced with increasing restrictions to water and land access. This has served to stimulate resentments and social grievances among the local resource users. Both Kenya and Uganda have strengthened their national institutional framework despite the ongoing processes of decentralization and the conclusion of devolution. At the national and economic level, the economic agendas have played a key role in terms of assuring natural resource access for economic actors. In this regard, the component water management authorities (e.g. Basin Water Resource Committees (BWRCs) and the Catchment Management Organizations) did not play a key role. Instead, the Ministry of Water and Sanitation and the Ministry of Water and Environment facilitate the implementation of the national economic agendas in the lake basins. The tightened legal frameworks which define the rights and codes to natural resource access, institutional decision-making and economic revenue sharing are still incipient, generally due to the lack of political willingness (inefficient bureaucracies and endemic corruption) and the national economic orientation. In the current study, comparing Lake Naivasha with Lake Wamala demonstrated that the broader institutional, regulatory and legal framework play the most important role in shaping the overall eruption of water-related conflicts at the local level. An unanticipated finding was the low level of co-ordination and co-operation amongst the local stakeholders. Instead, recurrent violent incidents to access and use natural resources are a major impediment to security and adaptive capacity.

6.3 Policy Recommendations

There are multiple constraints which result in increased tensions, and which enhance the potential for conflict and contribute, therefore, to an increased human insecurity. If those constraints are not mitigated, they might intensify already existing grievances and cause regional instability in the medium term. The study provided a bigger set of significant constraints which are associated with the eruption of water-related conflicts between the local stakeholders: (i) weak state institutions and institutional capacity, (ii) suffocating economic expansion and (iii) high level of distrust among the local actors and the motivation of revenge, particularly for Kenya. Given the constraints and blockades identified, in the following, three key entry points for co-operation instead of conflict are proposed: (I) changing the narrative; (II) a holistic approach to economic development and (III) empowering local stakeholders.

6.3.1 Changing the Narrative

The aforementioned analysis on water-related conflicts, vulnerability and governance within the two lake basins indicated that there is the need for a different narrative focusing on shared problems and, therewith, shared solutions. This suggests 'to look for mutual gains whenever possible . . . and the result [will] be based on fair standards independent of the will of the other side' (Fisher and Ury 2012: xxvi). It further enables a fair solution and protects the other party 'taken advantage of [one's own] fairness' (Ibid.). Another advantage of a changed narrative is its relation to an integrative approach. The latter can be defined as 'parties whose strategy is to create value [and] understand that they can get more [from a co-operation] than they would away from it' (Bercovitch and Jackson 2001: 73). To provide more solid, shared and jointly accepted solutions on all levels of decision-making, a comprehensive approach is required. This should focus 'on basic interests, mutually satisfying options, and fair standards' (Fisher and Ury 2012: 11, 14–15).

Separate the People from the Problem

Conflict participants are often emotionally involved with the conflict (e.g. anger or frustration) which can impede coming to an agreement, particularly when some kind of relationship between the parties pre-existed, as in this conflict. Therefore, the problem has to be disentangled from the people (Fisher and Ury 2012: 21–23). The challenge revolves around the narrative of economic development at the national level and access to natural resources including, primarily, water and land at the local level. However, access to water and land is relevant for social and economic stability and to preserve the integrity of the ecosystem. Resulting thereof, the political leadership is required to overcome their currently existing diplomatic conflict at the national and sub-national level of decision-making. Thus, both the national and sub-national governments need to feel equally involved in the process of natural resource governance. Nevertheless, the current institutional decision-making suppresses any further calls for co-operation. Hence, to address this issue, the root cause of the diplomatic conflict needs to be focused on. This does not mean a regime change, but it requires some major reforms of the existing political system and the political culture (i.e. corruption). Resulting thereof, a possible reintroduction of decentralization and devolution is seen as one way of mitigation against the institutional ineffectiveness. Dialogue among the political stakeholders and the catchment management representatives is a precondition to overcome overlapping areas of responsibilities, financial constraints or the formulation of legal and policy reforms, especially. The political institutions at a sub-national level of decision-making are weak, though they have the potential to perform far better with more commitment to transparency, accountability and performance orientation. Thereby, it might be easier to focus the attention on the current political polarization and problem and reduce the potential for future conflicts.

Focus on Interests, Not Positions

Although leadership is important to solve the diplomatic conflict, the water-related conflict between the local stakeholders needs to be addressed separately. The communities, small-holder farmers and pastoralists' interest are the actual cause of the dispute, not their positions (see Table 6.1). As these positions form the basis of any agreement, they must be prioritized. To achieve this, the underlying interests of both primary conflict actors need to be made explicit. This can happen by asking the other 'Why?' – why do the other side persist on their positions (Fisher and Ury 2012: 42–46, 50). By making the positions explicit and visible, like during joint community meetings, and thereby writing them down, it should be easier not to relapse into positional bargaining again. Furthermore, joint community meetings can be an essential platform to identify trade-offs and opportunities to promote sustainable agriculture and livelihood opportunities and safeguard security between the different local stakeholder groups. Lastly, community meetings help to build or to restore mutual trust between the different local stakeholders and to simplify broader dialogue and burden sharing. The onion analysis pointed to the shared interest of access to water and land, as well as social and economic developments. Specifically, the shared interest in improved water and land access illustrates a possible zone of agreement and depicts the most important controversial issue which therefore should be addressed first.

Invent Options for Mutual Gain

In order to find possible options from which both the political and local stakeholders' benefit, a communal meeting between the stakeholders and district or county representatives can be conducted which shall be free from any judgement or criticism (Fisher and Ury 2012: 62–65). Such a community meeting took place during the stay at Kikandwa community. In such meetings, various ideas may come up on how to solve the problem of land and water access, the usage of more sustainable farming techniques or security issues from which the viable ones can later be separated. In addition, both the primary, secondary and tertiary actors' interests will be taken into account and treated equally. Further, the idea of a fixed pie for everybody has to be overcome (Fisher and Ury 2012: 72). The national and sub-national decision-makers need to understand that their interests in economic development and institutional transformation can only be solved mutually. This becomes increasingly important regarding the revenues from the economic sites, the infrastructure and its likely potential for economic and financial wellbeing (Johnson 2010: 41). Regarding the local communities, it is the reverse, as they are not primarily interested in the revenues but in grazing and agricultural lands, improved water access and the set-up of educational and health facilities. Hence, the political stakeholders and the communities are interested in different objectives (Fisher and Ury 2012: 58–59).

Insist on Using Objective Criteria

Basing a possible agreement on objective criteria enhances the changes of a commitment (Fisher and Ury 2012: 84). Regarding the first narrative of economic development and institutional ineffectiveness, it might be advisable to look at both issues at the same time, as any economic transformation will, in the end, go hand in hand with a political transformation and effective institutions. Considering the question of revenue sharing of the economic investments undertaken and the use of the financial infrastructure, agreements between the different companies and the countries should be taken as objective criteria to find a fair solution which corresponds to the international standard and is also in line with the already-existing legal framework for revenue sharing. Thus, the concerns of the communities in matters related to improved educational and health care facilities might become second-ranked by making explicit the actual interests of the actors.

6.3.2 A Holistic Approach to Economic Development

Notwithstanding the earlier comments about changing the narrative, economically it will be essential to first regain control over the spiralling debt stock and to go beyond the credo of 'infrastructure first' and 'foreign investment benefit Uganda' (State House 2015). Both Museveni's and Kenyatta's understanding of economic development includes primarily infrastructure (with a focus on roads and transport) and pivotal trade partnerships with foreign investors, primarily Chinese and Indian ones. However, this economic understanding undermines a more holistic approach to development. A critical reassessment of the economic objectives and the change of the structural guidelines of the existing economic potential seems a requirement. Hence, the government needs to shift its directions to focus trimming the bloated public services (e.g. inflated administration and patronage network) and to focus on the set up of solid and fundamental public services (including health and education as well as improving basic sanitation and water supply) instead. To reach the desired levels of productivity and competitiveness, as laid out in the national development plans, both countries cannot work around the demographic challenge. Some community members highlighted that youth unemployment is a major obstacle to security in the basins. Thus, providing opportunities for the young generation will be key to a successful and sustainable development, especially in rural areas. Moreover, this could help to improve the security situation.

Especially around Lake Naivasha, the economic sites are already in place. Thus, a modernized and strengthened educational system that focuses on both critical thinking and creativity, as well as vocational training and technical skills, provides the young generation around Lake Naivasha with the opportunity to find well-skilled employment. Considering that more than 50 per cent of the population are still employed in the agricultural sector, education, especially in rural areas, should not only prioritize economic development but also address sustainability and natural resource management. Additionally, the labour market in both countries needs to be expanded so that it can absorb more young people. Thus, modernizing agriculture and drafting a more robust industrialization strategy shall contribute towards a greater productivity.

As indicated earlier, both presidents focus on foreign investments to improve the economic situation. Both Kenya and Uganda borrow money from foreign investors or foreign investors are hired for prestigious infrastructure projects or to implement the economic agenda. As a result, the debt levels of both countries are growing. There is concern that both countries are heading into a debt crisis, even though the debt levels can be considered sustainable currently.[7] It would be advisable that the governments critically assess the conditions under which they borrow money from foreign investors. The countries should find sustainable ways to both fund the economic transformation as well as include their country's own labour force instead of allowing the foreign investors to bring in their own. Possible first steps could include a joint engagement with the foreign partners which is in line with the five-year and long-term development plans. Secondly, the countries could attract foreign investments and should strengthen a domestic revenue generation without relying on unconditional loans from China or India. The former also requires the willingness of international organizations to rest and assess their growth assumptions for both economic expansion and monetary benefits. The international organizations' allowances towards the agricultural sector should not be based entirely on strengthening the agricultural sector but should also take into account the sector's vulnerability to external shocks and climate change.

6.3.3 *Empowerment of Local Stakeholders*

It can reasonably be argued that water-related conflicts between the local stakeholders are a result of their low adaptive capacity to cope with the different externalities.

Therefore, key entry points should include the empowerment of local stakeholders in matters related to water supply and water accessibility. This shall reduce the risk of outbreak, or the fear of outbreak of water-related conflicts in the medium to long term.

In both lake basins, local and non-governmental international and national actors try to achieve similar political, economic, as well as environmental aims. Local NGOs especially should be given more responsibilities in matters related to formulating, implementing and monitoring decision-making environmental processes concerning Lake Naivasha and Lake Wamala basin. However, to be capable of acting, they need to be better equipped financially, as well as have to be treated as a reliable and equivalent partner. Furthermore, the existing legal framework regarding a sustainable and equal handling of water resources needs to be implemented and put into action. Therefore, to reduce water induced conflicts in the basins, a more active, free and meaningful participation of all water resource users is inevitable. A good water and natural resource governance approach is aimed at including responsibility, transparency, equity and efficiency. In the end, it needs to be community-oriented overall. It should further include sanctions in case the laws are not followed as well as to hold all stakeholders accountable in case of natural resource and policy mismanagement.

Local resource users need to be empowered taking action to improve their water supply and accessibility instead of 'waiting' for the government to take action. Thus, local resource users can also put efforts to increase their own adaptative capacity. In the case of Lake Naivasha, most local resource users reside within one of four informal settlements within the closer surrounding of the lakes. However, these settlements lack basic amenities. As the instalment of water services are often costly and also will remain costly in the near future, a group-oriented approach is recommended. This group-oriented approach shall cover all local resource user groups. It might enable them to construct a water infrastructure which serves all their priorities while simultaneously reducing the costs for all of them. Furthermore, a co-operative sharing of resources would result in more advanced communication between the different resource user groups and thus reduce the levels of distrust. As proposed earlier, a community-centred approach also includes regularly held meetings to bring all local stakeholders together. During these meetings, employees from international and national NGOs in co-operation with local stakeholders shall provide the stakeholder groups with trainings and workshops on how to manage and use natural resources sustainably, especially in times of unpredictable weather changes. These meetings can further help to build trust and confidence among the different resource users. However, it is not sufficient to arrange community meetings only. To increase local low-key conflict prevention and resolution mechanisms, traditional institutions as well as peace meetings facilitated by NGOs and government actors need to be promoted. A failure to mitigate the already-existing conflicts will further add fuel to the fire and increase the vulnerability of the affected communities. This then might lead in the near future to regional destabilization. In summary, the local stakeholders need to address the Elephant in the room and work towards containing the land and water situation and to push towards a joint solution instead of following the credo 'every man for himself, god for us all'.

6.4 Brief Summary

Kenya's and Uganda's lack of state capacity has been a challenge in reaching equal access opportunities to water resources at the local level. Due to its direct interaction with social and economic factors at the national level and indirect interaction with climate change, domestically, increasing temperatures coupled with governance challenges and social and economic implications will affect multidimensional security, not only among the local stakeholders but in the medium to long term. Water and food security, livelihood challenges, internal migration and intensifying communal conflict due to impacts on water access and the continuation of traditional livelihoods are all issues that might exacerbate already existing grievances and lead to social tensions in the first place. Furthermore, communal conflicts and perceptions of insecurity might also be compounded by climate change impacts in the near to medium term.

The study showed that the competition for scarce water resources, which are actually still abundant in the lake basins, are central to violent and non-violent conflicts witnessed in the area. The eruption and persistence of those conflicts is an indication of a weak national institutional framework. The implementation of the existing governance structures is carried in disregard to the local stakeholders' abilities and means of livelihood. Thus, ill-informed and ill-planned political and economic interventions neither address the root causes of the problem nor meet the local stakeholder's level of knowledge. Moreover, inadequate policies focus on the complex tenure issues in traditional settings. Therefore, the development of a macro-economic agenda that leads to an economic transformation has to meet, first, the countries' citizens' abilities and, second, has to work closely within the existing political framework. Furthermore, dialogue between the national and sub-national political stakeholders is a precondition, especially to overcome the current political polarization. Independently, community participation in resource and conflict management is crucial for achieving a sustainable conflict-free resource sharing. However, national and international actors need to acknowledge the local stakeholder's high levels of vulnerability to water access which is attended by violent and non-violent behaviour. Thus, nothing of the aforementioned can be accomplished by silencing dissent.

6.5 Book's Conclusion and Avenues for Future Research

Water-related security risks are increasingly compounding. Because of existing political, economic and social challenges worldwide, it becomes more and more important to understand the impact of institutional and economic factors on human livelihoods. This book, therefore, has sought to answer the main research question 'How does water management influence innerstate low-key conflict intensity in

Kenya and Uganda?' by looking at and discussing the eruption of water-related conflicts at Lake Naivasha and Lake Wamala.

The research into water-related conflicts at water abundant areas is still in an early stage but, nevertheless, it is possible to highlight some points. Even in areas which do not seem water scarce from the outside due to their vulnerability to climate change, it does not automatically imply that conflicts over water are inexistent. When economic and political stakeholders, with different motivations and capabilities to influence national economic and political processes, weak institutions, marginalization and unequal resource access and distribution at play, local stakeholders fuelled the intention to forcefully pursue their right to water. In this case, water is more likely to contribute to violent conflicts, independent of the amount of water available.

Several overarching conclusions can be drawn and, therewith, highlight the contribution of this research for a better understanding of water-related conflicts in water-abundant areas. Firstly, the complex relations between water availability and conflict can only be understood if questions of vulnerability and adaptation for all identified stakeholder groups are answered. Consequently, any attempts to mitigate water-related conflicts need to address the specific vulnerabilities of each individual conflict actor.

This leads to the second point that the lack of political willingness to allow sub-national participation in resource policies leads to ineffective local adaptation processes, and therewith, the eruption of local-level conflicts. The localization of the institutional conflicts is more than a mere struggle in the study countries. In many countries of the Global South, issues of governance and economic upgrading is an arena for struggles over authority between the central state and 'twilight institutions', such as sub-national governments and customary institutions. This is especially the case in countries with weak state authorities which are, furthermore, highly vulnerable to being affected by climatic changes and transitioning from least developed to middle-income countries. The context is appreciated by international and national business actors where conventional donors have lost clout as a result of the growth in developing economies and the lack of a clear-cut system of responsibilities. It allows them to engage in national decision-making processes and to influence national policies in matters related to their interests.

Thirdly, this study has demonstrated that it is possible to grasp the complexity of the research objective by applying multiple research methods and theoretical frameworks. The comparative case study analysis allowed specific conclusions to be drawn about the causes and dynamics of water-related conflicts beyond the aspect of climate change. As climate change has led to the drying up of water bodies in arid areas, people are moving to humid and, therewith, water-rich areas. Thus, regions and areas that have not been affected by climate change yet will be affected by water shortages. Particularly, with respect to future research on the water and conflict nexus, the combination of the various methods and approaches used, as done in this book, might allow for a more in-depth qualitative field research with quantitative data analysis and modelling beyond the cases studied.

From this, the fourth point emerges, the increased need to focus on water abundant areas even though they seem calm on the surface. As indicated in Chapter 3, there is a need to establish a water and resource available georeferenced spatio-temporal database to researchers and decision makers to better understand the spatio-temporal distribution of water and economically relevant resources. This database should capture local and non-state conflict dynamics as well as allow to understand how the different levels of decision-making influence operations on the ground. With gradual development of databases and frameworks, it will be possible to test hypothesis on the relation of water and conflict in all areas, like the one developed here, in other conflict contexts.

Consequently, a relevant area for future research efforts would be to clarify how powerful international companies wield leverage in and across economic value chains and what implications this has on not only the shift in the end markets towards the south and the growing importance of securing access to water and land but also the supply in the agri-food sector for local and national actors. Another important area for future research would be to examine the implications of international cash flows to limit the effects of climate change on the available resources. Kenya and Uganda are dependent on the economic undertakings at Lake Naivasha and Lake Wamala. The COVID-19 pandemic resulted in the temporary disruption of global trade. Thus, an area for future research might be the analysis of the correlation of the economic effects of the corona crisis on water-related conflicts.

Notes

1 Bonafide is the purchaser of the land.
2 Kibanja is a person who settles on land with the consent of the landowner, the Bonafide.
3 Kabaka means king.
4 At the landing sites and in the informal settlements, there hardly exist latrines. Consequently, the residents have to go to the bushes and in the lake for basic sanitation.
5 In 2020, Kenya's debt stock was 69 per cent of its GDP. In 2020, Uganda's debt stock amounted 45.7 per cent of its GDP (IMF 2021).
6 For example, economic stakeholder versus economic stakeholder; economic stakeholder versus political stakeholder; political stakeholder versus political stakeholder; local stakeholder versus political stakeholder or farmer versus herder.
7 For Uganda, the external debt for the public and private sectors increased from 1 per cent to 41.4 per cent of the GDP in 2017–2018. At the same time, the debt owed to China is steadily increasing, from 3.3 per cent in 2011 to 20.3 per cent in 2017 (IMF 2021). For Kenya, the external debt for the public and private sectors increased by 5 per cent to 60.2 per cent in 2017–2018. At the same time, the debt owed to China is also steadily increasing, standing at 21.3 per cent in 2018 (IMF 2021).

Bibliography

BBC (2017): *No Bed of Roses: The Kenyan Flower Pickers Fighting Sexual Harassment*. At: https://www.bbc.com/news/business-39103419 [retrieved last 6 February 2020].

BBC (2018): *Kenyatta Wants Legacy to Be Fight Against Corruption*. At: www.bbc.com/news/av/world-africa-45343956/kenyatta-wants-legacy-to-be-fight-against-corruption [retrieved last 31 March 2020].

Bercovitch, J. and Jackson, R. (2001): Negotiation or mediation? An exploration of factors affecting the choice of conflict management in international conflict. *Negotiation Journal on the Process of Dispute Settlement* 17 (1), 59–77.

Donald, P. F., Sanderson, F. J., Burfield, I. J. and van Bommel, F. P. J. (2006): Further evidence of continent-wide impacts of agricultural intensification on European farmland birds, 1990–2000. *Agriculture, Ecosystems & Environment* 116 (3–4), 189–196.

Fisher, R. and Ury, W. (2012): *Getting to Yes. Negotiating an Agreement Without Giving in*. Business Books. London.

Government of Kenya (2010): *The Constitution of Kenya*. Republic of Kenya. Nairobi.

International Monetary Fund (2021): *External Debt*. At: http://databank.worldbank.org/data/reports.aspx?source=2&series=DT.DOD.DECT.CD [retrieved last 24 April 2021].

Johnson, D. H. (2010): Border battle line. *International Journal of African Renaissance Studies* 5 (1), 36–47.

The Observer (2019): *Understanding Mailo Land and Dual Ownership*. At: https://observer.ug/viewpoint/59858-understanding-mailo-land-and-dual-ownership [retrieved last 24 April 2020].

The Republic of Uganda (1998): *The Land Act*. Government of Uganda. Kampala.

The Statehouse of Uganda (2015): *Why Uganda Is the Best Investment Location in Africa*. At: www.statehouse.go.ug/why-uganda-best-investment-location-africa [retrieved last 24 April 2020].

Vivekananda, J. (2015): *Peace Audit Kenya*. International Alert. London.

Zerga, B. (2015): Ecological impacts of Eucalyptus plantation in Eza Wereda, Ethiopia. *International Invention Journal of Agricultural and Soil Science* 3 (4), 47–51.

INDEX

For Product Safety Concerns and Information please contact our EU representative GPSR@taylorandfrancis.com Taylor & Francis Verlag GmbH, Kaufingerstraße 24, 80331 München, Germany